OPTIMAL YIELD MANAGEMENT

To
my wife Shoshana
and my children Ron, Michal, Irit and Tamar
with love

Optimal Yield Management

Edited by
DAN RYMON
Department of R & D Economics
Agricultural Research Organization
Israel

Avebury
Aldershot · Brookfield USA · Hong Kong · Singapore · Sydney

Published by

Avebury

Gower Publishing Company Limited
Gower House
Croft Road
Aldershot
Hants GU11 3HR
England

Gower Publishing Company
Old Post Road
Brookfield
Vermont 05036
USA

British Library Cataloguing in Publication Data
Optimal yield management.
1. Agriculture. Efficiency
I. Rymon, Dan, 1940-
630

Library of Congress Cataloging-in-Publication Data
Optimal Yield Management.
Includes index.
1. Crop yields--Congresses. 2. Farm management--
Congresses. I. Rymon, Dan, 1940-
SB16.067 1988 631 88-7435

ISBN 0 566 05707 7

Printed and bound in Great Britain by
Antony Rowe Limited, Chippenham, Wiltshire

Contents

Contributors

Reimar von Alvensleben, University of Hannover, Hannover, F.R. Germany
Shawki Barghouti, The World Bank, Washington D.C., USA
Bnayahu Bar-Yosef, Agricultural Research Organization, Bet-Dagan, Israel
Avner Carmi, Agricultural Research Organization, Bet-Dagan, Israel
Ariel Dinar, Agricultural Research Organization, Bet-Dagan, Israel
Gideon Fishelson, Tel-Aviv University, Tel-Aviv, Israel
Chanan Geisenberg, Retired, Extension Service, Ministry of Agriculture, Israel
Arieh Goldman, The Hebrew University, Jerusalem, Israel
John R. Hanks, Utah State University, Logan, Utah, USA
Lynn Hinton, Cambridge University, Cambridge, United Kingdom
Oded Hochman, Ben-Gurion University, Beer-Sheva, Israel
Chaim Jogev, Extension Service, Ministry of Agriculture, Israel
Eliezer Kletter, Extension Service, Ministry of Agriculture, Israel
Guy le Moigne, The World Bank, Washington D.C., USA
Avishalom Marani, The Hebrew University, Faculty of Agriculture, Rehovot, Israel
Dan Marom, Extension Service, Ministry of Agriculture, Israel
Uri Or, Kibbutz Magal, Israel
Zvi Plaut, Agricultural Research Organization, Bet-Dagan, Israel
Philip V. Rasmussen, Utah State University, Logan, Utah, USA
Uri Regev, Ben-Gurion University, Beer-Sheva, Israel
Sam Roth, Cotton Production and Marketing Board, Tel-Aviv, Israel
Dan Rymon, Agricultural Research Organization, Bet-Dagan, Israel
Yair Sachs, Extension Service, Ministry of Agriculture, Israel

Ezra Sadan, The Hebrew University, Faculty of Agriculture, Rehovot, Israel
Yoav Sarig, Agricultural Research Organization, Bet-Dagan, Israel
Ido Seginer, Technion - Israel Institute of Technology, Haifa, Israel
Hovav Talpaz, Agricultural Research Organization, Bet-Dagan, Israel
Eugene D. Ungar, Agricultural Research Organization, Bet-Dagan, Israel
Dan Wallach, INRA, Toulouse, France

Preface

In this book we try to emphasise several of the principles on how to manage the optimisation of yields and to elaborate these principles by analysis of empirical evidence.

The book is selective rather than encyclopedic in nature. The experience of approximately the last decade indicates that optimising yields is an interdisciplinary issue, and it constitutes the backbone of this book. The probability that an achievement in a single expertise or discipline will have a significant impact on the sector (i.e., agriculture) or even on a single branch (i.e., growing cotton, raising dairy cattle) is low. However, the book does not present anything of a revolutionary character in the theory of each of the professions taking part in its composition.

This book contains two main ideas: first, the interdisciplinary viewpoint and the indispensable need to cooperate efforts of different professionals in order to obtain optimal yields, and second the knowledge that optimisation is in most cases distinct from maximisation. By organising the team taking part in this compendium we bring to light several innovative works, as an outcome of the overall agricultural technology developed in Israel and its yield achievements.

In recent years, much progress has been made in implementing these principles into systems analysis and real-life management practices. By providing the broadest sketch of the involved disciplines, we shed light on the subject of how an optimal yield is reached from various angles. We have found from experience that professionals better understand what is written by their colleagues than what is written by the rest of the expertises which participate in the optimisation process; it is probably a result of extreme specialisation and the large amount of newly produced 'data'. Binding the 'mix' together challenges the desire for a hybrid effect through increasing the probability of interaction; by that it contributes to the process of optimising yields, as compared to the mere summation of all of the components.

The *introduction* to this book is dedicated to the concept on how to manage toward an optimal yield through enlarging and modernising its scope. It broadens the meaning of the 'output' this managerial effort is aimed at optimising, as well as the 'inputs' sharing this process.

The *first part* is devoted to generally related considerations and experience: the organisational structure, the marketing innovative and analytical ability, and the experience in transferring the developed knowledge to the farmers who finally decide whether it is worthwhile.

Under the assumption of the marketing orientation of agricultural activities, the *second part* is devoted to an investigation of the market potential and the foreseen trends, as a guide to the management of optimal supply to meet the demand and the improved determination of an agricultural policy.

As every decision process starts from an acquaintance with the state-of-art, this approach was followed with regard to the optimisation of yields. It was decided to concentrate the demonstrations within the framework of field crops, especially cotton and vegetables, with special focus on tomatoes. The *third part* of the book is devoted to the background history and developments, the difficulties and the achievements, and the considerations on how to optimise the yields.

The *fourth part* consists of seven chapters examining the role and ways of optimising the use of inputs. It starts with the contribution of the production factors which are commonly *not* analysed: radiation, the plant's root zone and agricultural engineering. This is followed by a presentation of the influence on yield management of more conventional inputs (e.g., water, fertilisers, etc.), but with a slightly different approach in relation to the output: obtaining a better quality, a more precise control on timing of supply or a saving of water per production unit.

The *fifth part* includes descriptions and discussions on the role of modelling in the optimisation of several production decision processes, and economic analyses for the sake of better resource allocation; all these are reviewed in close relation to the management of optimal yields.

This book is composed of chapters that were gathered by the organising committee of the First International Symposium on Optimal Yield Management held on the occasion of AGRITECH 86, the agricultural exhibition of Israel, and were presented in this symposium, under chairmanship of the Deputy-Minister of Agriculture, Member of Knesset

Avraham Katz-Oz. This collection was initiated and designed - from the planning stage onward - so as to be later incorporated into a complete book. The symposium drew together more than thirty scientists of natural sciences, social sciences, engineering and management. It was the honour of this editor to chair the organising committee, to take command of the collection of the contributions, and after the symposium, to assume responsibility for editing and preparing it for press.

Any errors or omissions in this book are entirely the responsibility of the editor. However, an exceptionally large number of people contributed to this book, some of whom are mentioned below. Members of the organising committee of the symposium deserve credit for providing their assistance since the initiation stage; of these Mr. Rafi Glick, of the Israel Export Institute, Dr. Yoav Sarig, Dr. Baruch Retig and Prof. Joseph Shalhevet, all of the Agricultural Research Organisation (The Volcani Centre), and Prof. Ezra Sadan, of the Faculty of Agriculture (The Hebrew University, Jerusalem) extended the most support. An important role in the quality of the contributions to this book is based on comments we have collected from a series of knowledgeable referees, whose names are not detailed here, and to whom we would like to express our thanks. The organisational work was furnished by ORTRA Ltd. (Israel), with the special dedication of its director general Mr. Zvi Gelfand jointly with Mrs. Tamar Carmeli, and I wish to thank them for their invaluable efforts.

The symposium was attended by close to 120 participants from 19 countries. There were two days of lectures and discussions and a one-day field trip, where the issues of optimising yields were demonstrated. I feel obligated to thank Mr. Joseph Elkana, of the Extension Service in the Ministry of Agriculture, Mr. Uri Or, of Kibbutz Magal, and Mr. Roni Nakar, of the Agricultural Research Organisation, for their assistance throughout the stages from the preparations till the end of the symposium.

Several Israeli companies and organisations contributed their financial assistance, which made possible the running of the symposium, and it is a pleasure to thank them on behalf of the organising committee for their generosity: Israel Chemicals Ltd. (ICL); Israel Export Institute; Israel Vegetables' Production and Marketing Board; Omni Agro-industries Ltd.; Pery - Development and Application Ltd.; and the Vegetables' Growers Association.

I would like to thank Mrs. Diane Greenberg for her invaluable editorial assistance, in preparing and proofing the manuscripts to the stage that this book is brought to the reader, Ms. Sarit Shalhevet for preparing the index and assuming responsibility for proofing, Ms. Talila Michaeli for her skillful typing and Gower Publishing Co. for making the publication of this book possible.

Dan Rymon
Bet-Dagan, ISRAEL

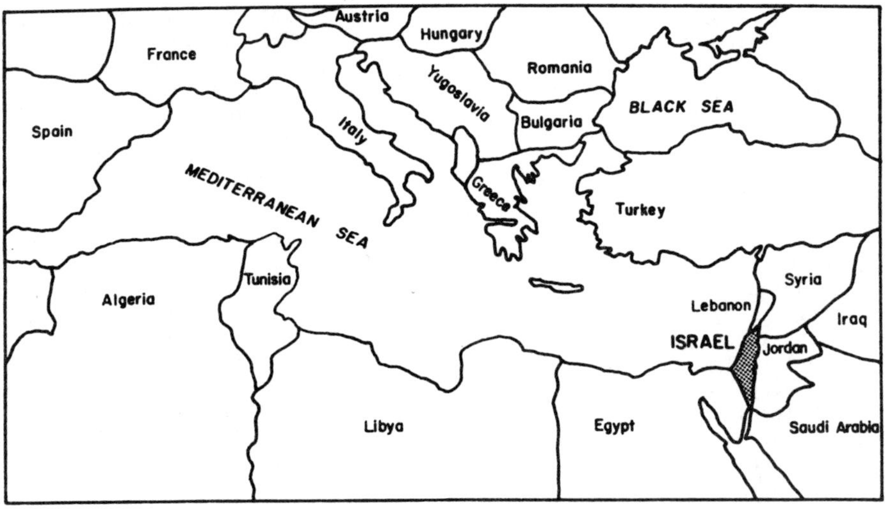

Map of the Mediterranean Countries

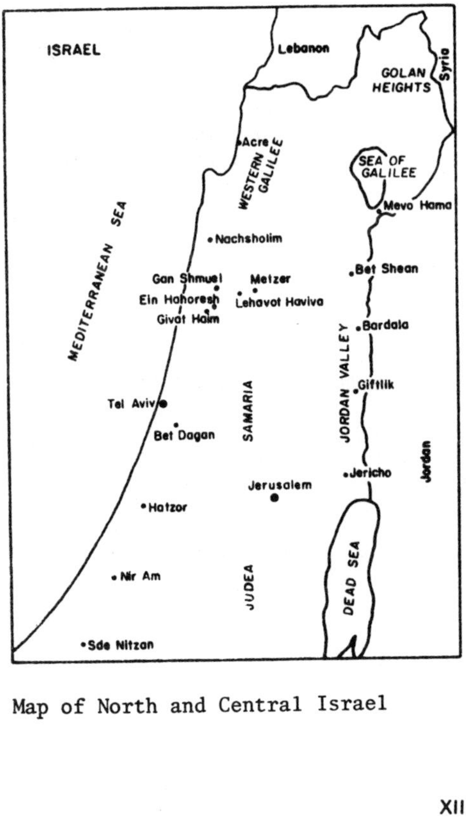

Map of North and Central Israel

INTRODUCTION

1 Conceptual thoughts on what are optimal yields

DAN RYMON

Modern agriculture is, more than even in the past, a business; this characteristic of the sector influences the decisions taken by farmers, in which profits and success are a major driving force. This statement does not override complementary considerations, those of the farmer and those of the state in relation to the farmers and to the entire agricultural or the whole rural community. For the sake of this discussion, however, these issues are set aside and will be dealt with separately. Our attention focuses on the main pragmatic thoughts of the decision makers in the farming sector. Within this framework, management to optimise yields in its broadest and commonly understood sense is certainly a dominant managerial issue. This chapter argues that yield management should be the main activity of both the farmer and the state, even though in the traditional theory on optimisation of yields it does not appear in such an explicit manner. This viewpoint is even more detailed and broadly demonstrated in subsequent chapters of this book.

* The author is grateful to Ariel Dinar, Gideon Fishelson and Ezra Sadan for their valuable comments on an earlier draft of this chapter. However, the opinions and errors are entirely those of the author.

The basic concept of optimal economic behaviour of a producer is considered through an analysis of a production function; this is defined as a relation between a series of production factors, or inputs, and one or more outputs (product(s) and/or service(s)). Production function analysis is only briefly elaborated here and readers are referred to the appropriate textbooks for a fuller treatment of the subject.

Following, for example, Henderson and Quandt (1958, chapter 3), the traditionally known optimisation process assumes a production function that states the outputs (q) - within a certain and known technology - to be a function of the quantities of the variable production factors (inputs) $(X_1, X_2, \ldots, X_n)$, and one or more fixed inputs $(X_{n+1}, \ldots, X_m)$:

$$q = f(X_1, X_2, \ldots, X_n / X_{n+1}, \ldots, X_m) \quad (1)$$

This is assumed to be a single-valued continuous function, defined for non-negative values of inputs and output levels and with continuous first- and second-order derivatives. Additional clarifications and detailed assumptions can be found in economics textbooks.

A graphical description of the complete production function deserves a space of m+1 dimensions; a two dimensional description, as found in every textbook (e.g., in Redman and Redman, 1981, chapters 2 through 4), of one output and one variable input, will suffice to present our main issues (see Figure 1).

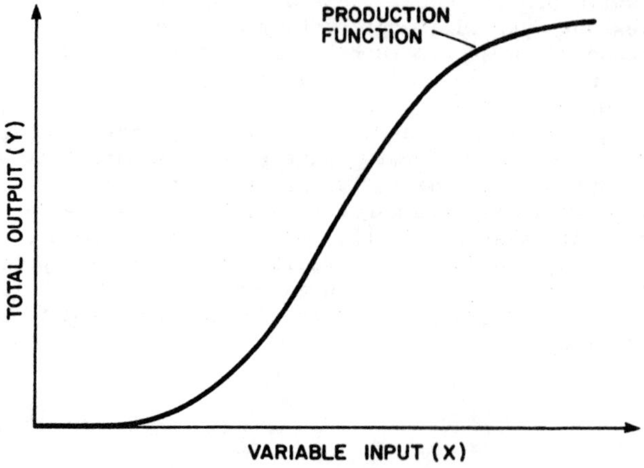

Figure 1 The classical production function

In this description, the production curve divides the area into the feasible outputs originated from each amount of the variable input - together with the fixed input(s) - and the rest of the outputs which are unattainable from these quantities of inputs. A point on the curve denotes getting the maximal output out of every defined combination of the variable input and the fixed input(s). This 'being on the curve' in itself is a managerial accomplishment, and one can verify first if this has been achieved as a guidance for a physically (or technologically) optimal yield. This is a negative way to distinguish between efficient cases and inefficiencies: a nonoptimal yield is obtained if one can increase the yield without a change or an increase in the inputs. Among the technologically optimal yields there is still a need to find the best, or the economically optimal one.

Optimisation of output(s) from the standpoint of the producer (i.e., profit maximisation, at least in the long-run and in its broad sense) is a maximisation of profits subject to constraints put by the technology and the markets (those of the inputs and those of the outputs). The latter means making an allowance for prices of inputs and outputs. The first-order conditions for optimisation in the production factors' market are that the values of the marginal products derived for each production factor is equal to that factor's price (PX_n). (The marginal product of each input X_n is denoted by MP_n and defined, following Ostrosky and Koch (1979), as the change in the total output that occurs when an additional unit of the input X_n is employed; its 'value', the VMP_n, is calculated by multiplying X_n by the product's price P.)

$$VMP_1 = PX_1, \; VMP_2 = PX_2, \; \ldots, \; VMP_n = PX_n \qquad (2)$$

In a similar way, we have a first-order condition in the market where the output(s) is sold, arguing for an equation of the marginal revenue (which is defined - following Ostrosky and Koch (1979) as the change in total revenue that occurs when the firm sells an additional unit, and denoted as MR. In certain cases it is equal to the price of the product, P) and the marginal cost (which is, following Ostrosky and Koch, the change in the total cost that occurs when an additional unit of output is produced, and denoted as MC), broadly defined.

$$MR = MC \qquad (3)$$

Similarly to the previous description of the graphical presentation of the production curve, these two optimisation considerations (see Figures 2 and 3) are a search of the best point on each of the curves, where the price ratio line is tangent to the production curve (equivalent to the point where the VMP_n equals P_n), and where the marginal cost equals the marginal revenue.

Returning to the formal conditions leads to second-order conditions; it is required that the principal minors of the relevant bordered Hessian determinant be positive. In a less formal way, this phrase means that profit must decrease with respect to further application of either X_1, X_2, and so on until X_n, or the MPs of all inputs are decreasing.

Another analysis of the optimisation issue is given by Varian (1978); after a detailed, formal and abstract analysis he concludes:

"the level of output should be chosen so that the production of one more unit of output should produce a marginal revenue exactly equal to its marginal cost of production. Another decision of the firm is to determine how much of a specific factor - say labour - to hire. The fundamental condition for profit maximisation tells us that the firm should hire an amount of labour such that the marginal revenue from employing one more unit of labour should be equal to the marginal cost of hiring that additional unit of labour".

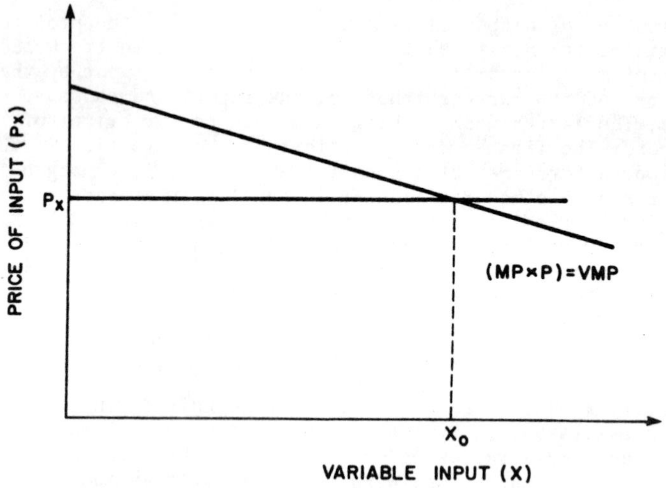

Figure 2 The market for a production factor

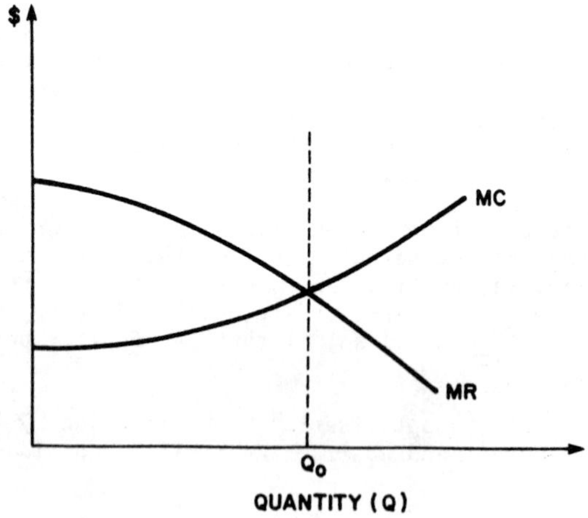

Figure 3 Marginal revenue and marginal cost

Since this evaluation of production economics and management was given, several changes have occurred. They might be seen as deviations from the pure traditional framework of the theory; they constitute important features of the real world and, if not included now, they should be incorporated in the theory:

1. Estimation procedures, or econometrics, presents the departure from the previously described production curve, which assumed that in each input level the curve determines the best and unique performance. It allows for accidentally exceeding or falling below the, supposedly, best possible value. It therefore sees the production possibilities as limited within a certain range around this value, without loosing the justification to be called value, without losing the justification to be called optimal.
2. Programming, operations research, and modelling of the production phenomena. These techniques break away from the assumed continuity. By this the employment of several optimisation methods appears feasible, and very significant achievements, especially in the understanding of interrelationships between variables, have been gained. These two extensions might be viewed as the best attainable proxis to the traditional economic theory.
3. Policy determination, market developments, training and extension work, are considerations on the dynamics of the production process, as compared to the previous comparative statics. The production curve undergoes continuous changes influencing the determination of the optimal yield, further to the mentioned above. The meaning of an optimal yield becomes dynamic, or time dependent; and comprises the dynamics caused by man on-farm, and the environment around the production activity.
4. Enlarging the scope, to incorporate contributions from related disciplines, like sociology, anthropology, organisations and management, marketing, finance etc.

The last two directions are the main streams in which this book is designed to contribute, and this chapter aims at defining them explicitly.

The concept of the production function potentially has an extension option on both its aspects, 'the output' and 'the input'. The functional relation between the two, which is left for a future study, are more of a technical nature.

In addition to the simple notion of an output in a market characterised by competition as a homogeneous product, composed of well-defined identical units (see, for example, chapter 3 in Leftwich, 1976), it is worthwhile bearing in mind the following considerations:

1. *Quality of the Product*. Different qualities of the 'same' product frequently originate from the same production process. Looking at this output distribution as a series of distinct products, or, alternatively, referring to the monetary value of the output, offers no satisfactory explanation of the following market phenomenon: the existence of

different qualities of the same product. Tomatoes of different perceived quality, as an example, are sold in a given market for a price difference of up to three-fold. No cost difference or another economic variable can explain this phenomenon, except the superior quality for which the consumers are willing to pay (see chapter 3, 6 and 7). For a theoretical understanding of this price differential, the reader's attention is drawn to Zusman (1967); his analysis gives an in-depth explanation of the consumer's demand rationale, that lies behind the grading and sorting schemes. He does not refer to the supply side, namely the distribution of the qualities originating from a production process, which is the focus of our interest here. The third facet of the quality issue is the optimisation of inputs in the production: the implications of the decision to produce a certain quality (say superior quality tomatoes), or a mix of qualities, will be discussed later.

2. *Timing of the supply*. Supplying a product to the market in different dates, of the week, of the year, or of the product lifespan, changes completely the 'environment' in which the acceptance and the price of this product are determined. An extreme case occurs if one supplies a known product to a market vacant from any supply of it (i.e., providing fresh peaches, nectarines etc. to European cities from late March to mid-April, namely before Easter); the marketing agency can demand almost any desired price, within the bounds of fruit prices. Theoretically, one can imagine and create a daily market, but an experienced economist or a marketing expert knows how impractical this idea is (chapters 7 and 8 elaborate this issue for two European markets).

3. *Branding of a product*. There is much evidence on the difference in profitability to the producer between 'commodities' and branded products. The marketing literature deals with this issue under the framework of 'brand loyalty', whereas an economist will interpret it as an entirely or partially monopolistic market position. We are probably facing a trend of increasing the market share of the branded and the semi-branded products; moreover, even for products sold traditionally in huge quantities as commodities (see chapters 3 and 7) there is a move toward stronger identification to enable the consumer to express his loyalty through repeat buying. (Supermarket chains, for example, identify the source of their supply even for products such as vegetables.)

4. *Reputation of a producer/supplier*. This feature may have roots in the brand sold by the producer (or supplier), but this is not a unique explanation, and the producer's/supplier's reputation may be an important characteristic as well. Constantly offering reliable products, or an especially good quality control, or a diversified year-round supply etc. can build his reputation. These and other features contribute to one's market position (see chapter 7), and it may differentiate one product from another (even if there is no objectively identified difference). From this moment onward this position is of economic value; moreover, if a producer has acquired such a position, he can classify his output to the category that harnesses this reputation and the other category - the rest of the produce - can be sold outside the reputed channels.

5. *Promoted and advertised product*. These investments in the marketing of a product pay at the final determination of market clearing conditions, namely the equilibrium prices and qualities.

Most of these considerations might be presented as inputs, by looking at their production process (i.e., the investment in quality control, or an advertising campaign etc.), instead of presenting them at the 'output' side of the production function of the finally sold good or service. It remains a dilemma how to define and which signs to attach to these issues, but there are no doubts related to their acceptability.

On the arguments' side of the production function, defining which are the relevant 'inputs', the basic theory deserves several extensions. Without dealing with its theoretical depth one may find a partial analysis of this type within the framework of 'input quality' and its significance. Several recent works indicate the additional gains a producer should expect in terms of yield, when input quality varies; among these are works related to quality differentials of the input of land, such as that of Offutt *et al.* (1987) and Casewell and Zilberman (1986). The description which follows is different in nature, and refers to the variables rather than to quantification of their impact:

1. *Inclusion of normative variables* in the equation is an alternative to adding constraints to the analysis of the optimal product-mix. An example may be the decision to avoid the hiring of labour from outside the farming community, with the knowledge that it reduces the net income. A possible outcome of such a position may be the development of a machine to cope with the needs it imposes (i.e., see chapter 13), a machine that otherwise would not be developed. From the state's point of view, a priority may be set in accordance with a social or a global set of norms; cases which demonstrate it are the settling of farmers on a near-border line, despite the relatively less favourable agro-economic conditions, or the answer to the question "what society are we looking for in the rural areas?".

2. *Timing in the performance of the agricultural activities*. This straight-forward phrase is probably the clue to the distinction between developed and underdeveloped farming: fulfilling the 'regular' tasks (of irrigation, spraying of biocides, harvesting, etc.) at the optimal time differentiates those who may expect to optimise yields from the rest of the farmers. (See chapters 14, 15, 18 and 19.)

The inclusion of a cost item in the production equation does not refer to whether it was implemented at the optimal time, the answer to which is quite complex.

3. *Radiation* is the means by which plants acquire energy for their entire production activities. This important production factor has an optimal level as do other variable inputs (see chapter 12). It is safe to assume that in most specified production functions this variable is not specified explicitly, but is probably hidden or indirectly represented by another radiation-influenced variable.

4. *Information* is an input in the sense that it plays a role in the managerial capability to decide, and its purchase is an alternative to that of other inputs. As an input in a production process it clearly

ranges from the minimal degree to the maximum with an optimal level in between.

5. *Organisations* surround the production activity and participate in the possibility of achieving an optimal yield: these include credit suppliers, inputs' buying cooperatives, marketing agencies of different types and others. Their potential contribution, constructive or destructive, starts with the question of their deficiency or existence, and continues with their operational effectiveness, as well as issues of their monopolistic degree in the production and marketing system (see chapter 2).

6. *Keeping pace with the dynamics in the use of inputs* is a multi-facet consideration to include:

6.1 *Development of technological options* to remove constraints to the efficient operation of the production process. In some cases the economics of an industry in a certain location, region or country, depends on the development of a new machine (see chapter 13). In technical terms it means a distinct production function, and the optimal producer's income is changed.
6.2 *Transfer of technology* is a production factor in optimising the yields of a region or a farming community compared with an analysis of a single farm (see chapter 5). The performance of the producers is a quantitative variable like most other inputs: therefore the role of the smoothness of the technology transfer is important: its success is connected to overall production increase or improvement, quantitatively or quality wise.
6.3 *Dynamic adaptive marketing* is an important activity in optimising the economics of a production process. Surviving or even benefiting from the dynamics in both the preferences of the consumers and the developments in the marketing organisations - as presented, for example, in chapters 7 and 8 - deserves a marketing system that is easily adopting itself to the market events. It may be difficult to define specifically such a system as an input; however, it is possible to specify a non-parametric variable of it as an input. In chapters 2, 3, and 6 one may find examples of dynamic thinking about several aspects which can lead to the building of such a system.

7. *Management* is the clue for optimisation of yields. It is composed, firstly, of optimised single activities as described, for example, in chapters 18 and 19, and, secondly, from the overall managerial analysis and decision making. The latter is especially difficult: (a) it is preferred to start a development action by technological improvements rather than by reforming social structures (as can be learnt from the experience described in chapters 4 and 5), (b) in analysing a new technology, emphasising its contribution in monetary terms is preferred over the practice of presenting the improvement in technical terms, such as per unit of area (see chapter 20), and (c) is preferred to transfer the optimal messages, incentives and signals to attune the producer, on behalf of an industry or a state, with the objective to produce in accordance with its interest, quality requirements, reputation, etc. (see chapters 6 and 21).

8. *Entrepreneurial dedication* plays a major role in the success of production activity to perform optimally. As in the previous cases this is a non-parametric variable.

These considerations, and others that were not detialed here, comprise an interdisciplinary discussion of how to optimise the yield. An important point, in conclusion, is the simultaneous nature of both the decision taken and the achievement of the optimum. The management of how to optimise yields requires taking all the above into consideration, in the daily activity of a manager responsible for that task, to whom this book should conbtribute, both on the general and the on data supply level.

REFERENCES

Casewell, M. and Zilberman, D., (1986), 'The effects of well depth and land quality on the choice of irrigation technology', *Amer. Jour. Agr. Econo.*, 68: 798 - 811.

Henderson, J.M. and Quandt, R.E., (1958), *Microeconomic Theory*, New York, McGraw-Hill Inc.

Leftwich, R.H., (1976), *The Price System and Resource Allocation*, Sixth edition, The Dryden Press.

Offutt, S.E., Garcia, P. and Pinar, M., (1987). 'The distribution of gains from technological advance when input quality varies, *Amer. Jour. Agr. Econ.*, 69(2): 321 - 327.

Ostrosky, A.L., Jr. and Koch, J.V., (1979), *Introduction to Mathematical Economics*, Boston, Houghton Mifflin Co.

Redman, B.J. and Redman, J.C., (1981), *Microeconomics: Resource Allocation and Price Theory*, Westport, Connecticut, AVI Publishing Co.

Varian, H.R., (1978), *Microeconomic Analysis*, New York, W.W. Norton & Co.

Zusman, P., (1967), 'A theoretical basis for determination of grading and sorting schemes', *Jour. of Farm Economics* (previous name of the *Amer. Jour. Agr. Econ.)*, 49(1): 89 - 106.

PART I
MANAGING TOWARD OPTIMAL YIELDS – OVERALL CONSIDERATIONS

2 Effective crop management, input deficiencies and underlying social organisation

EZRA SADAN

Cropping, or the production process, taking place in an experimental station could be looked upon as a set of mere technological relationships. Real-world cropping involves more than this and so does real-world crop management. In order to clarify this point, we inquire into input deficiency as a common phenomenon of poor management.

The following sections delve in some detail into fertiliser and irrigation water deficiency. By linking these phenomena to capital market failure and the lack of patterns in farm cooperation, an attempt is made to illustrate certain corrective strategies aimed at eliminating input deficiencies and allowing for effective crop management.

MATERIAL AND COMPOSITE INPUT DEFICIENCY

Proper crop management concerns appropriate input-output relations. However, inputs cannot be identified for that purpose as mere material entities, nor can cropping or production be classified as purely technological processes.

A simple notion of improvements in crop management is often associated with a simple perception of input deficiencies. Deficiency of a particular input is supposedly reflected in marginal benefits exceeding the respective cost. Corrective programs, intended to augment these deficiencies, seem particularly attractive where shortage is substantial, and the input in question has a critical effect on production levels of a strategic crop (Palmer, 1972). But even a simple strategy, aimed at a single (critical) input and one particular (strategic) crop, requires certain changes in the social organisation supporting agricultural production.

For instance, consider the case of an Andean smallholder peasant growing potatoes and utilising a (nitrogen) fertiliser. That fertiliser is considered 'under-utilised' where the value of the marginal product exceeds the input price. With reference to the open-market spot-prices of the output and input in question, under-utilisation or input deficiency is a common phenomenon among peasant smallholders. The question is: what is the real cause of the apparent discrepancy between input prices and the respective values of the marginal product?

In order to answer this question,iit is necessary to recognise the difference between the material input and the composite input consisting of the fertiliser itself, and certain associated services. The latter include the handling, hauling, distributing and financial transactions involving the material input.

Apparent under-utilisation of material inputs is often traced to inadequacies affecting the non-physical components of the composite input. In rural Latin America, for instance, credit appears to be the truly deficient component. Because credit is rationed, the cost of utilising fertilisers for a smallholder potato-farmer is far higher than the respective social opportunity cost. Also, the value of the marginal product does not correspond to the spot-price of the potatoes registered in the wholesale market. Hence, from the peasant's point of view fertilisers are not being under-utilised: because of the excessively high real interest rates a wide discrepancy between the social benefit and the respective social cost coexist with an equality between private benefits and costs. That equality implies inputs levels approximating zero where exorbitant interest rates result in prohibitive costs.

INPUTS DEFICIENCY AND SOCIAL ORGANISATION

At the root of the material input deficiency lies a scarcity of the truly deficient component. The latter is related to a lopsided distribution of influence over rural economic affairs. Control over marketing and credit in the countryside, concentrated in a few hands, is an outcome of the global political structure (de Janvry, 1981). At the same time the realisation of monopolistic practices and capital rationing are particularly related to the smallness of the farm enterprise.

Related to the case of the smallholder Andean potato grower, the input intensity in three Andean states (Peru, Equador and Bolivia) is comparable with that of a neighbouring relatively developed country (Chile):

	Relative fertiliser consumption (1971 - 73) per ha %	Relative number of tractors (1974) per ha %
Chile	100	100
Peru	49	39
Equador	22	11
Bolivia	6	4

These ratios were calculated from a study of twenty-five years in the Agriculture of Latin America (1950 - 1975) prepared by CEPAL (1978). The authors of this study point out that despite the intensified participation of the international and national public sector in the financing of the farming sector, agriculture is still suffering from capital rationing. They indicate that the role of institutional credit varies considerably from one country to another.

Particularly sensitive are countries suffering from minifundismo (smallness). Lopsided power-structure and minifundismo are reflected among other things in a dependence upon a single dominant source of farm credit. Often that source is a transporter serving as a roving agent of a dominating wholesaler situated in the metropolitan center. That particular agent is the peasant's only connection with the potato market, on the one hand, and the capital market, on the other. Credit penetrated down from the wholesaler reaches the peasant through payments advanced at real rates exceeding 100%. For example, payments advanced at the outset of a six-month season amount to 50% of the price eventually due at the farm gate. That price in its turn equals 75% of the price realised by independent farms. All in all, the dependent smallholder pays 50% interest on a six-month loan, that is, an annual rate of 125%! Similar rates result from credit advanced from retailers of seeds, fertilisers and grocery.

In either case, transporters and retailers possess an exceptionally accurate knowledge of the financial affairs of their dependent clients. They ration their supplies accordingly and extract monopolistic gains through exorbitant rates of interest.

Apart from enriching the intermediaries and their masters, the excessively high rates contribute to long- and short-term deficiencies. The prevalent deficiencies are reflected in a simple comparison of the number of tractors and fertilisers in the Andean states, on the one hand, and in Chile, on the other hand. Institutional credit is better developed in Chile than in the neighbouring countries.

Chile is one of the better organised countries in this respect. Nevertheless, even in Chile the relative weight of non-institutional credit (extended by retailers) has been quite heavy at least at the outset of the period analysed by CEPAL (Corsson, 1970). The rates realised in that non-institutional segment of the capital 'market' were as exorbitant as those calculated above for intertemporal transactions in the potato-belt in the Andes (Nisbet, 1967). In other words, a comparison between the Andean states and Chile weighs the heavily deficient against a mildly deficient country.

A STRATEGY DIRECTED AT THE ELIMINATION OF INPUT DEFICIENCIES

Personal relationships characterise credit transactions of the farm sector in the High Income Countries (HIC). Lenders in the HIC know their clients financial affairs as well as their non-institutional counterparts in the developing countries (DC). However, in the HIC the knowledgeable operator is an officer of a credit organisation (a bank, a credit union etc.) which is an integral part of the country's institutional credit market. A family farm in a HIC is a relatively small enterprise; nevertheless, the turnover of that small business enterprise is large enough to pay for personalised services of the lending institutions. Furthermore, in the HIC an unsatisfied family farmer can sever relationships with one credit institution and shop around for an alternative one. This transfer is a costly exercise and a farmer is bound to avoid transfers as long as he can. However, in a community of family farmers even a small number of transfers will suffice to demonstrate a potential, and thus help maintain a spirit of independence and competition. Under these circumstances the system is effective in providing for steady flows of loanable funds to finance the farm business. It is efficient as it maintains rates of interest consistent with the 'prime' rates charged in the case of large incorporated, well-established borrowers. The differences between rates charged on credit extended by a family farm and that benchmark are properly attributable to cost differentials due to a greater risk and a costlier operation.

In the DC, smallholder peasants are truly small. At the going commercial rates and commissions their business will not justify the attention of a decision-making officer of a credit institution. Small-scale and isolation reduce the range of potential intermediaries between the peasant and the markets, in general, and the sources of institutional credit, in particular. Under these circumstances, the traditional intermediaries may or may not be potentially efficient agents of the 'capital market'. In practice they take advantage of their monopolistic-monopsonistic position. Hence, rates charged, implicitly or explicitly, on intertemporal transactions are exorbitant. The discrepancy between these rates and the ones free of monopolistic gains creates a distributive disparity, deficiencies and waste.

A strategy aimed at the improvement of crop management through the reduction of input deficiencies in DC must overcome the current as well as the historical results of smallness and isolation. In this respect, even the least ambitious strategy involves a certain degree of social reform.

SIMPLE AND COMPOSITE STRATEGIES

The simple example contemplated so far focuses on an apparent mismanagement involving a single input and a single crop. Needless to say that in a far-reaching strategy incorporating several inputs, 'new' and 'old and several crops, 'new' and 'old' may require far-reaching measures of social reform. For instance, limited and provisional measures may suffice to overcome, at least temporarily, a deficiency in a single input applied in the production of an established crop. But, an attempt to improve crop management through the introduction of new varieties or new crops and a (balanced) combination of various old and new inputs will most probably require a permanent reform.

In a sense, a limited strategy is nested in the more comprehensive one (Hopper, 1968). Similarly, the far-reaching strategy combines the reform requirements of the constituent simple strategies. A necessary condition for the accomplishment of an effective crop management through a comprehensive strategy is a transformed farm-supporting organisation, capable of catering to a wide array of new demands. Transformation would require a removal of obstacles hindering the flow of credit. For the reason mentioned above, and for other reasons, the uprooting of these impediments must be particularly radical if access to credit as well as other channels is to be universal, including the small peasantry.

ADDITIONAL ASPECTS OF INPUT DEFICIENCY AND 'MISMANAGEMENT'

So far the discussion of the relationship between social structures and economic efficiency in agriculture focused on the organisation of the farm supporting system, visualised as an arm of the respective national institutions (markets). However, it may be pointed out that there is a question of the organisation within. The need to organise the farmers themselves in order to allow for effective input allocation is particularly visible in the case of irrigation water (Bromley *et al.*, 1980). This case manifests strong and weak points in the capacity of a farming community to allow for the reconciliation of competition for self interest and the need to cooperate (Thorenton, 1985). Interdependencies in the case of irrigation water require group action. In its absence the resource may be misused and depleted. Allocation may be ineffective resulting in waste, on the one hand, and input deficiencies, on the other.

Organising the smallholders in "users' associations", etc., may be difficult, as it may rely on the ability to solidify revised norms of cooperation or even new ethics.

A CONCLUDING REMARK

Poor crop management in developing countries is likely to reflect failures in the farm supporting system and the pattern of farm-cooperation. Corrective measures aimed at these failures are rather difficult because they involve indispensable elements of social reform and, perhaps, cultural change. At any rate, cropping and production cannot be looked upon as if they were mere technological processes. Crop management is more than a technical manipulation of material inputs and outputs.

REFERENCES

Bromley, D.W., Taylor, D.C. and Parker, B.E., (1980), 'Water reform and economic development, institutional aspects of water management', *Economic Development and Cultural Change*, 28: 265 - 387.

CEPAL, (1978), *25 Anos en la Agricultura de America Latina, Rasgos Principales (1950 - 1975)*, Santiago de Chile, CEPAL/ILPES.

Corsson, P.R., (1970), *Agricultural Development and Productivity Lessons from the Chilean Experience*, Baltimore, Resources for the Future, John Hopkins University Press.

Hopper, D.W., (1968), 'Investment in agriculture: the essentials for payoffs', in *Strategy for the Conquest of Hunger*, New York, The Rockefeller Foundation.

Janvry de, A., (1981), *The Agrarian Question and Reformism in Latin America*, Baltimore, John Hopkins University Press.

Nisbet, C., (1967), 'Interest rates and imperfect competition in the informal credit market of rural chile', *Economic Development and Cultural Change*, 16: 73 - 90.

Palmer, I., (1972), *Science and Agricultural Production*, Genova, UN Research Institute for Social Development.

Thorenton, D., (1985), 'Organising the use of irrigation water', *Agricultural Administration*, 17: 168 - 175.

3 Tastes research – a major tool in the development of new agricultural products

ARIEH GOLDMAN

INTRODUCTION

New products play a very important role in the packaged consumer goods industry. Many of the items we currently see on the shelves of food stores did not exist a few years ago.

The introduction of new products involves a major commitment of resources. In addition to the direct costs of research, development and marketing, opportunity costs of time and management attention are also involved.

A large number of studies have documented the fact that the risks of failure of new consumer products are very high (Urban and Hauser, 1980; Wind, 1982). Studies analysing the experience of various consumer goods companies have found that only one out of 20 new product ideas actively considered become a successful product on the market.

How can one reduce the risks of market failure of new products? The accepted answer is the institutionalisation of the process of new product introduction and systematic market analysis and testing. It is believed that if the process is accompanied by a set of appropriate market tests, we will be able to detect, early-on, potential failures. Also, getting consumer reactions to the product and generating diagnostic information about the strengths and weaknesses of the product when it is still in the development stage will also allow the developer to modify and increase its chances of success.

The agricultural sector is far behind the packaged goods sector with regard to introduction of new products and sophistication of this process. However, the increased interest shown by consumers in fresh food products and the fact that supermarket chains now regard agricultural produce departments as central to their competitive posture and as major contributors to sales and profits has resulted in the increased emphasis on new products. It can be expected that, as in the packaged goods industry, only a few of the new products will be truly new in the sense that they were not available before (*Time*, 1986) while most new products will be improved versions of existing items. In any case, we are still in the early stages of this movement into new products in the agricultural sector. Indeed, most R & D efforts in this area continue to be directed toward 'process' type product improvements, such as higher yields, longer shelf life, increasing ease of handling etc., rather than consumer-oriented product improvements yielding product changes that are aimed at a better fit with consumer preferences.

The taste element is of special interest in the development of new agricultural products. Many of the R & D efforts are aimed at developing better-tasting products. Unfortunately, the established taste-testing procedures fall short of the needs of the developers of new agricultural products. There is a need to develop such procedures that will make it possible to evaluate consumers' responses and preferences for differently tasting versions of new agricultural products.

This paper briefly describes a behavioral taste evaluation procedure developed specifically for use in agricultural products. During the last few years the procedure was applied to a number of cases of new agricultural products (e.g., tomatoes, grapefruits, easy-peelers, melons and mangos). In the first section a short account of taste-testing procedures is given, followed by a description of the potential contribution of taste research to the development of new agricultural products with examples of several types of outputs the procedure yields. A more detailed discussion of various aspects of this procedure has been given by Givon and Goldman (1984a, 1984b, 1985 and 1987).

TASTE TESTING

Two features characterise the current taste-testing practices in the food industry (Amerine *et al.*, 1965; Moskowitz, 1983; Moskowitz, 1985). First, many tests are conducted on small (a dozen or so members), pretrained, in-house panels. The reliance on panels stems from the belief that, because of the tasting-task complexity, regular, untrained consumers will find it difficult to detect differences or state their true preferences. Many firms believe that by training their panels, they avoid the need to directly assess the members' discrimination ability each time they perform a taste test. Second, the emphasis has been on perceptual tasks, such as 'difference testing', rather than on the evaluation of taste preferences. Consequently, the issue of evaluating consumers' discrimination ability in the area of preferences, has received little attention in the literature and by sensory evaluation professionals. The reason is to be found in the preoccupation with the small panel, and in the conviction that it is difficult to generate meaningful taste information from regular, untrained consumers. While for some purposes, such as product development and quality control, that involve perceptual discrimination tasks, the small trained panel is an adequate tool, preference information generated from such panels is not very meaningful. The evaluation of product acceptability, and how a new version is likely to fare against competitors, requires an analysis of the preferences of large, representative, consumer samples. Recent developments in the food industry, such as the increased emphasis on taste segmentation and on product positioning on the basis of taste (Buchanan, 1985), makes the measurement of taste preferences even more important. Yet, if taste tests are to be administered to large consumer samples rather than to small pretrained panels, the measurement of the discrimination ability of each consumer should become an integral part of the preference measurement procedure.

The procedure for measuring taste preferences developed here makes it possible to evaluate the discrimination of each respondent from the preference data he generates. Also, it overcomes many of the other problems, such as lack of attribute-based information that is needed for diagnostic purposes of the various product versions.

GRAPEFRUIT STUDY

The procedure will be described in the context of a study evaluating a number of varieties of grapefruits.

The grapefruit study was intended to test consumers' preferences for, and evaluation of, grapefruits originating from different regions of the country and characterised by different levels on the standard 'maturity' tests. Five varieties of grapefruits were tested, some Israeli and the others from countries which compete against Israeli grapefruits in European markets.

TASTE-STUDY PROCEDURE

The grapefruit study was conducted in France and Germany. A sample of consumers who were characterised as regular users of grapefruits were recruited and brought to a study suite in the center of town. The study in each of the countries was conducted in two major cities. On entering the test room, the respondent was seated at a table facing the interviewer. Five plates were placed on the table, each containing half of a grapefruit. Each plate was identified with a two-digit code, seen from the interviewer's side. A pitcher of mineral water, and a plate of crackers were placed on the table, and respondents were encouraged to rinse their mouth, eat a piece of cracker and rest, 'whenever they feel they are losing their ability to taste'. The tasting sessions lasted an average of 20 minutes. Since we know from earlier studies that grapefruits are consumed in two major ways: eating and in juice, we also conducted a test in which grapefruit juice was tasted.

The test tasks were designed to measure the preferences of each respondent, using two methods: paired comparison and rank ordering. Information about ideal points was also collected and socioeconomic and consumption data for checking for possible segmentation.

The sequence of experimental tasks was as follows:

1. *Paired comparisons:* the five study grapefruits yielded ten combinations of all possible pairs. A respondent was presented with each pair by the interviewer, according to a predetermined sequence, tasted both grapefruits and indicated his preferred grapefruit.
2. *Dollar-metric:* after determining the preferred grapefruit in a pair, the respondent was asked, 'Let's assume the price of a grapefruit in the market is now X (reference price was determined based on the actual situation in the market), by how much money should the price of the least preferred grapefruit go down so that you will prefer both in the same way'. The purpose of this question is to evaluate the intensity of preferences in money terms.
3. *Preference ranking:* respondents tasted all five product varieties and arranged them according to their preferences indicating the most preferred, second, etc.
4. *Consumption and buying habits:* at this point in the interview schedule, respondents answered a set of questions about their grapefruit (or the item tested) consumption patterns and their buying habits.
5. *Product attribute rating:* respondents evaluated after additional tasting, the position of each of the five study products, on a scale (we used a five-point one). For example, where the attribute sweetness was concerned the scale went from 'very sweet' to 'not sweet at all'. In the case of grapefruits, we used the attributes to describe the concept of 'taste of a grapefruit'. These attributes were generated from the literature, in depth interviews with people in research, industry and consumers. They were pretested in a prior grapefruit taste experiment.
6. *Ideal point:* respondents were asked to think about their main usage of grapefruits and then to indicate the desired amount of each of the attributes in their 'ideal' product for their main usage. The scale used was a five-point scale. Then they were asked to indicate how important it is for them that the ideal grapefruit should contain this amount of the attribute.
7. *Socioeconomic information:* respondents answered questions regarding their socioeconomic status and were debriefed.

STUDY RESULTS

We will give some of the typical results generated by the above taste-testing procedure, and discuss preferences and ideal products.

Preferences

Preferences for each of the study products, grapefruits in this case, can be assessed either directly based on the paired comparison task, or by direct ordering or indirectly by looking at distance from ideal points. Here we use direct ordering. In Tables 1 and 2, the preferences for the five study grapefruits are shown. In Table 1, tasting involved eating the fruit and in Table 2 it involved drinking the juice.

Table 1

Preferences (rank ordering) for grapefruits - eating, in % (France)

Grape-fruit	*first*	*second*	*third*	*fourth*	*fifth*	*first and second*	*fourth and fifth*	*Characterisation of grapefruit*
	R	*A*	*N*	*K*				
26	25.6	16.5	18.2	27.3	12.4	42.1	39.7	splitting
42	13.2	14.0	9.1	20.7	43.0	27.2	63.7	dominated
53	23.1	20.7	19.0	20.7	16.5	43.8	37.2	splitting
61	22.3	24.0	21.5	16.5	15.7	46.3	32.2	domineering
84	15.7	25.6	31.4	14.0	13.2	40.1	27.2	domineering

Table 2

Preferences (rank ordering) for grapefruit juice, in % (France)

Grape-fruit	*first*	*second*	*third*	*fourth*	*fifth*	*first and second*	*fourth and fifth*	*characterisation of grapefruit*
	R	*A*	*N*	*K*				
26	28.3	17.5	17.5	24.2	12.5	45.8	36.7	domineering
42	3.3	15.0	15.0	23.3	38.3	18.3	61.6	dominated
53	11.7	16.7	31.7	14.2	25.8	28.4	40.0	dominated
61	30.8	27.5	19.2	17.5	5.0	58.3	22.5	domineering
84	20.8	23.3	16.7	20.8	18.3	44.1	38.1	splitting

In the present survey, grapefruit no. 42 was clearly a dominated grapefruit in both the eating and juice drinking tests. In contrast, no. 61 had a domineering profile in both cases. Grapefruit no. 26 had a 'splitting' profile in eating and a better (domineering) one in juice. No. 84 had the same pattern but in reverse order. Finally, grapefruit no. 53 was found to have a splitting profile in eating and a dominated one in juice. Interestingly enough, in France this grapefruit is considered to be the best in taste and fetches a premium price. One of the findings of this study is that some of the Israeli grapefruits (e.g., nos. 61 and 26) have a superior taste.

The preference data shows that the yellow Israeli grapefruits tested (nos.26 and 42) are heterogeneous in taste. Less taste heterogeneity was found in the two red grapefruits tested - nos. 61 and 84. In other words, as far as taste is concerned the yellow grapefruit category consists of two products each with a different taste quality.

When consumers choose directly between yellow and pink and red grapefruits, the market splits about 39% for yellow (nos. 26 and 42), 38% for red (nos. 61 and 84) and 23% for pink (no. 53).

The general pattern of taste preferences among the grapefruits is similar when they are evaluated for eating and for juice. However, two grapefruit types, nos. 42 and 53 fare much worse as juice than for eating.

Finally, there is no one grapefruit that 'fits' the French market. Even the 'worse' grapefruit finds a niche in this market, although its customers are smaller in number than for other types.

Intensity of preferences. The intensity of preferences for each grapefruit type was evaluated through the dollar-metric method. In general, the average intensity was between 20 - 30% of the reference price. That is consumers were ready to pay a premium of between 20 - 30% on the average for this preferred grapefruit type. Grapefruit no. 53 turned out to be in an especially strong position - those who preferred it, were ready to pay, on average, a 30% price premium for it.

Ideal grapefruit and product attribute profiles

The respondents evaluated each of the five grapefruits on 14 attributes. In addition they stated how much of each of these attributes they would like to see in the ideal grapefruit.

The ideal profile in Germany and in France appears in Table 3.

Table 3

Comparison of ideal grapefruit profiles in Germany and France
(averages on the scale 1-5, where 5 is the highest level)

	France	*Germany*
Sweetness	4.10*	3.42*
Acid and bitterness	2.01	2.04
Juiciness	4.63	4.66
Acidity	2.26	2.39
Aftertaste	2.07*	2.65*
Interesting taste	4.12*	4.60*
Sharpness of taste	1.88*	4.41*
Thickness of skin	4.14	4.10
Firmness	4.03*	3.56*
Gentleness of taste	4.31	4.35
External appearance	4.27	4.59
Internal appearance	4.48	4.76
Intensity of external colour	-	3.78
Intensity of internal colour	-	4.96

* Differences significant at the 0.05 level.

The profile of the ideal grapefruit in Germany is very different from that in France. The Germans prefer a grapefruit which is less sweet and with a 'stronger' taste (greater aftertaste and a more interesting and sharper taste). They prefer a somewhat less firm grapefruit.

It is clear that grapefruit taste is a complex phenomenon. Consumers do not want a grapefruit that is overly sweet, overly juicy or that totally lacks bitterness, acidity or thickness of skin. They want a grapefruit which is sweet but not very sweet, juicy but not extremely so, has some acidity, some bitterness and some aftertaste.

Attribute profile of the grapefruit

The strengths and weaknesses of each of the grapefruits can be evaluated by comparing the average score a grapefruit received for each attribute to the ideal level consumers wish to have for this attribute. A grapefruit that is nearest to the ideal level is a better one than a grapefruit that is more distant from the ideal.

In Table 4 we see the average deviation of each of the study grapefruits from the ideal level for each attribute. The squares indicate that the largest deviations from the ideal on each particular attribute and the circle in the smallest one.

Table 4

Average deviations from the ideal grapefruit (France)

	Grapefruit no. (types)				
Attributes	*(84)*	*(61)*	*(53)*	*(42)*	*(26)*
Sweetness	-1.08	-.38	-.16	-2.03	-.99
Bitterness	+1.06	+.66	+.18	+1.76	+.54
Juiciness	-.69	-.51	-.41	-.83	-.32
Acidity	+.85	+.26	-.08	+1.69	+.50
Aftertaste	+.73	.77	+.55	+1.04	+.37
Interesting taste	-.59	-.53	-.95	-.38	-.62
Sharpness of taste	+1.00	+.72	+.39	+1.66	+.53
Thickness of skin	-1.11	-.80	-1.07	-1.06	-.82
Firmness	-.52	-.61	-.65	-.51	-.63
Gentleness	-1.04	-.81	-1.08	-1.38	-.83
External appearance	-.51	-.31	-1.48	-.71	-.49
Internal appearance	-.71	-.50	-.97	-1.14	-.55
Easiness of eating	-.88	-.73	-.67	-1.05	-.57

A minus sign indicates a lower level than the ideal; a plus sign, a higher level.

Grapefruit no. 42 suffers from many weaknesses. It is farther away from the ideal in a very large number of attributes. It is not sweet enough, too bitter, not juicy enough, too acid and has too much aftertaste and sharpness of taste. It is not gentle enough in taste and its appearance is the worse. In contrast, grapefruits nos. 53 and 61 each has a number of attributes that constitute their strength (see circled scores). No. 53 for example is strong in 'internal' taste attributes such as sweetness, low acidity and bitterness, but is weak in external appearance and other physical attributes.

This type of analysis allows us to evaluate the strong and weak points of each grapefruit.

REFERENCES

Amerine, M.A., Pangborn, R.M. and Ressler, E.B., (1965), *Principles of Sensory Evaluation of Food,* Academic Press, New York.

Buchanan, B.S., (1985), 'Can you pass the comparative ad. challenge'? *Harvard Business Review,* July - August, 63:4, 106 - 113.

Givon, M. and Goldman, A., (1984a), *Preference Measurement and Consumer Taste Discrimination Ability,* Working Paper 806/84, The Israeli Institute of Business Research, Faculty of Management, Tel Aviv.

Givon, M. and Goldman, A., (1984b), *Taste Tests Perceptual and Preference Discrimination Ability,* Working Paper 830/84, The Israel Institute of Business Research, Faculty of Management, Tel Aviv.

Givon, M. and Goldman, A., (1985), *A Cross Cultural Study of the Taste Structure of a Food Item,* Working Paper 863/85, The Israel Institute of Business Research, Faculty of Management, Tel Aviv.

Givon, M. and Goldman, A., (1987), 'Perceptual and Preferential Discrimination Abilities in Taste Tests, *Journal of Applied Psychology.*

Moskowitz, H.R., (1983), *Product Testing and Sensory Evaluation of Foods,* Food and Nutrition Press Inc., Westport, Conn.

Moskowitz, H.R., (1985), *New Directions for Product Testing and Sensory Analysis of Foods,* Food and Nutrition Press Inc., Westport, Conn.

Time, (1986), 'A is for apple? No atemoya', August 11, 61 - 62.

Urban, G. and Hauser, J., (1980), *Design and Marketing of New Products,* Prentice-Hall.

Wind, J., (1982), *Product Policy,* Addison Wesley.

4 Irrigation management and sustaining agricultural yields

GUY LE MOIGNE AND SHAWKI BARGHOUTI

INTRODUCTION

The remarkable growth in agricultural production over the last several decades has been strongly associated with the expansion of irrigation and the advent of efficient production systems generated by 'The Green Revolution'. Investment in irrigation in developing countries has been successful in advancing agricultural production for increasing food and cash crops. This remarkable success has not been easy to achieve, however. Indeed, it will be more difficult and more costly to sustain or advance in the coming years, because of the declining return on agriculture, and because of the worsening economy of many developing countries. Many of the irrigation systems that have been built in the last three decades have not been well maintained. Substantial effort and investment would be required to rehabilitate these systems and to ensure proper operation and maintenance. Also, many of these systems require modernisation, incorporating efficient design and higher water control in order to address the evolving needs of modern agricultural technology, increasing cropping intensity and more flexible and diversified cropping systems. A major objective for future investment in irrigation development should be on advancing irrigation in Sub-Saharan Africa, where the benefits of irrigation have not yet been felt in many African countries that are suffering from declining food production and alarming deterioration in agricultural resources.

Yields of the main agricultural crops have increased during the last several decades. The advent of production technology largely generated by 'The Green Revolution' of the 1960s has produced a remarkable increase in the production of rice, wheat and maize. Improvement of other cereal and legume crops, and cash crops, is receiving increasing attention and is producing encouraging results in the arid zones of Africa and the Middle East. There has been remarkable improvement in agricultural production technology in several important countries in South and East Asia. India, for example, has become a showcase for increasing crop production. From being a food-deficit country in the 1960s, India achieved self-sufficiency in food grain in the mid-1970s. In the period 1950 - 1983 India's cereal production increased from about 55 to 140 million tons per year, and over 50% of that increase is attributable to irrigation. This achievement is a result of an agricultural strategy that emphasised the use of high-yielding varieties, development and expansion of irrigation, and increased fertiliser usage - all in selected areas of high potential for increased productivity.

Between 1950 and 1985, India's gross cropped area expanded by 30%, while the area under irrigation more than doubled in absolute terms from 22.6 million ha in 1950 to 49.6 million ha in 1985. Other countries have also made remarkable progress. Indonesia, for example, invested in expanding its irrigated areas and in the rehabilitation of existing irrigation systems. Rice production in Indonesia has increased by over 40% since 1960. Between 1965 and 1985 the Indonesian average rate of growth of rice production, area harvested and yield were 4.4%, 1.1% and 3.3% p.a., respectively. More recent rice increase has been spectacular. For example, between 1978 and 1982, the increase in production was 6.8% p.a. This growth trend over the last two decades has resulted in a steady rise in per-capita availability of rice in Indonesia from an average of 90.6 kg in 1960 - 1967 to over 140 kg at the present time. The direct impact of irrigation on increasing yields is confirmed in field data from India and Indonesia. Current yields of irrigated wheat in India average about 2.7 t/ha, while unirrigated wheat produces about 1.2 t/ha. Similarly, Indian irrigated rice averages 2.2 t/ha while unirrigated rice produced about 1.2 t/ha. In Indondsia, the wetland rice yields have risen steadily since 1960 and have averaged over 4 t/ha since 1982, although dryland yields are less than half this level. The yields of irrigated wheat and rice in the USA and Australia are almost double those achieved in Asia.

Other examples of remarkable achievements in improving agricultural production through irrigation are available from Thailand, the Philippines, Latin America and many other countries. Indeed, rapid expansion of irrigation throughout the world during the same period has increased in areas under irrigation from about 94 million ha in 1950 to an estimated 271 million ha in 1985, of which 75% is developing countries. Expansion of irrigation has been strongly associated with yield increases throughout the world. A recent report on world resources states that through the 1970s, as the world average annual production of grain increased by 2.6% a year, productivity gains per hectare accounted for two percentage points or about 80% of the growth. The report also states that, while the overall rate of growth in agricultural production for developing countries has been increasing - reaching 3.3% during 1977 - 1982 - the growth rate for the world as a whole has been declining. It is not clear whether this decline is because technological limits have been reached or it is simply the result of increasingly saturated market conditions compounded with increased cost of production.

CONSTRAINTS TO FUTURE EXPANSION IN AGRICULTURAL PRODUCTION

Future expansion of agricultural growth will become more difficult because of several technical, economic and environmental constraints that are likely to slow down the pace of growth or could even reduce improved yields that were achieved over the last three decades. At the technical level, expanding irrigation has not resulted in achieving maximum improvement in crop yields. Irrigation projects are built without much built-in support for modernising cropping systems. The engineering technology employed in these projects has in many cases not been matched by advanced cultivation technology. Thus, cost of crop production under irrigation tends to remain high because yields have not increased enough to compensate for the increasing use of a range of inputs including farm machinery, fertilisers, pesticides, herbicides and improved seeds. Future increases in agricultural productivity will be largely based on improving the performance of existing irrigation schemes and on increasing usage of inputs, which could be expensive and require advanced level of management and supervision. These requirements are unlikely to be met in many developing countries without further technical and financial assistance.

Data available on cultivable land around the world indicate that further increase in agricultural production is unlikely to come through the expansion of cultivation onto new lands, partly because the new lands may be in the wrong place. Areas of high population densities - where additional food supplies are needed - have the least amount of available land with high agricultural potential, and more than 60% of the population of the Third World live in countries rated land scarce. In addition, most unused cultivable land has its own technical problems. Central Africa's tropical rain forests, for example, lack infrastructure and are infested with a host of serious diseases that plague human and livestock populations. Large areas of Sub-Saharan Africa, where food deficit is growing to crises proportions, are too dry to cultivate. Similar examples could be listed in several Latin American countries where agricultural land cannot be expanded rapidly enough horizontally to produce additional food needed by the world's increasing population. Future expansion of irrigation to new lands is likely to be difficult and expensive. Most suitable sites for irrigation have already been developed, leaving marginal areas and less suitable locations for development at an average cost running at $ 5,000 - 10,000 per hectare for irrigation development. The average cost of irrigation development in South Asia in the 1970s was $ 1,200/ha.

The main economic factors affecting investments in irrigation are likely to be precipitated by the enormous increase in production of the main agricultural commodities. The production of rice, wheat, cotton, rubber and palm oil has in the mid-1980s reached levels so high that prices have plummeted below actual cost of production. India and Indonesia, countries that once faced widespread food shortages themselves, contributed 100,000 tons of wheat and rice, respectively, to drought stricken countries in Africa in 1985. The fact that Indonesia, which was until the 1980s the largest rice importer in the world, is now self sufficient in rice should be one of the important factors in the decline of rice prices. Global recession following the oil price increases of the late 1970s, the commodity surplus brought about by improving agricultural performance and weak economies have caused depressed worldwide demand for raw agricultural materials including rubber, coconut, palm oil, cotton as well as manufactured goods. Thailand is experiencing problems with its surplus rice, especially since American rice is highly subsidised for export purposes. Malaysia is now unable to market its palm oil. Previously interest in palm oil production in Malaysia was very strong because prices moved upward in the early 1970s (and reached $650/ton in early 1984). The price fell almost $ 190/ton in early 1986 when the biggest buyer, India, withdrew from the market, a new US Farm Bill subsidised cheaper soybean oil, a range of palm oil products was delivered from Indonesia and the production of Philippine coconut oil, a close substitute to palm oil, was increased. The decline in cotton prices have seriously affected Sudan and several countries which have invested heavily in irrigation for cotton production. The list of countries with surplus commodities could be extended. The situation, however, requires a careful look at what should be done to protect investments already made to advance agricultural production, especially in irrigation, drainage and water management.

Since its establishment after World War II, the World Bank has supported investment in irrigation and drainage projects. Indeed, the very first loan to a developing country made in 1948 amounting to $13.5 million was for a power and irrigation project. Since that time and up until 30 June, 1986, World Bank agricultural lending has amounted to almost 42 billion of which more than 14 billion was for some 355 irrigation projects. More than 80% of lending for irrigation was in the last decade. Irrigation and drainage lending represent 10% of all Bank lending and about 33% of agricultural sector lending. The Bank has perceived investment in irrigation as critical to expanding cultivated land and increasing yields of all irrigated crops. The return on investment in irrigation has been attractive and has encouraged further expansion because the value of commodities produced was relatively high, and the cost per hectare for irrigation development was relatively low. This is no longer the case. Costs of irrigation tend to rise with time, and current investments are required to finance more expensive system components, such as water storage facilities, and conveyance systems with built-in flexibility for water delivery to farm units to meet crop requirements. The recent values of irrigated commodities are not as attractive as before.

Rehabilitation and modernisation of existing irrigation system

Rapid expansion in irrigation systems, rising cost of irrigation development, increasing crop intensification, and rapid introduction of modern inputs have created new demands for specialised skills in operation, maintenance and management of irrigation systems.

Unfortunately, technological progress in these areas has not been introduced into many irrigated projects, with the result that many such recently-built projects in the developing countries have not been well maintained nor properly operated. Many schemes require immediate investment for rehabilitation and modernisation. Lack of good management and proper maintenance have reduced yields and increased crop losses. Inefficiency in irrigation management is visible in enormous losses, expanded waterlogged areas and silted canals. Salinisation is currently a problem in most irrigated areas in arid and semi-arid zones. In India alone, more than 7 million ha are affected by salinity and alkalinity. Further areas are degraded by waterlogging because of poor drainage and lack of modern irrigation and drainage design. The technical problems in land drainage (especially subsurface drainage) to prevent waterlogging and salinity at the root zone, have been addressed through advanced technology for horizontal or vertical drains. The enormous progress in horizontal drainage has been made possible with the advent of perforated corrugated plastic drainage tubing, and the introduction of lazer grade-control equipment, trenchless laying machines as well as by the use of powerful deep-drainage equipment which is capable of operating at higher speed and greater depth than the older trenchers. Vertical drainage is expanding to irrigated areas where there is an aquifer which allows good water movement from the root zone to the aquifer. When fresh water is obtained from the aquifer, the drainage effluent can be used to supplement the surface water supplies. The use of tubewell drainage is being adopted in Pakistan, Syria and northern India, and efforts are being made to expand this investment where possible. The main constraint to advancing drainage methods in developing countries is the cost that remains high when compared with the overall farm income. Many farmers who produce low-value irrigated cereal crops with limited income are unlikely to spend between $ 1,000 - 2,500, the average cost of a land drainage scheme. A more efficient method to reduce the problem of drainage is the improved irrigation application method, either through better water management including the conveyance and distribution systems, or through more efficient systems such as the center-pivot sprinkler, drip or bubbler irrigation, all requiring high capital investment.

The cost of modern irrigation schemes has become prohibitive when compared with the early Asian irrigation systems developed in proximity to flowing streams which required little more than simple diversion structures, usually of a temporary character. Management and distribution of water were also simple, mainly requiring limited attention to the temporary structure and diversion channels. Major investment has recently been made in constructing storage systems which could permit a considerable increase in cropping intensity, but require not only a higher level of capital investment but also a higher level of water management. Too many irrigation projects are designed merely to deliver water with little or no attention as to how the water is to be used in the field. The design of most of such systems is based on a rigid schedule of water delivery with little sensitivity to particular crop needs. Such systems impose on the cultivators several constraints that could limit crop flexibility and could lead to reduction in crop production or abuse of the irrigation system. Other projects are designed with fully regulated supply of water with the intension of maintaining equity of distribution among farmers according to timely and reliable schedules. Both rigid and flexible systems have not been properly maintained in many developing countries; adding further constraints to sustaining crop production.

Improvement of water management

These constraints have been well-described, and their impact on crop yields has been assessed through several studies. Some of the best data on this subject have come from a series of experiments conducted at the International Rice Research Institute (IRRI), which confirms the importance of water management and impact or lack of it on crop yields. The results of these experiments indicate that, for rice, the yield per unit of water declines with even moderate (early) water stress. Thus, it is exceedingly important to design irrigation systems which ensure reliable supply of water for rice. It is observed in many irrigated areas in rice-producing countries that farmers will go to great, even destructive, lengths to obtain adequate water for their crops from the irrigation system. (Rice farmers in some areas often damage farm turnouts in their attempt to secure adequate water from poorly operating irrigation systems.)

Modernising a water delivery system would require that irrigation components be designed to assure delivery of peak-period water requirements. They must provide adequate water, either in river flows or in man-made storage, and canals must operate at capacity large enough to carry the volumes needed (clean of weeds and silt) and must have adequate water control over a wide range of canal flow rates. These conditions have not been met in major irrigated areas of the world; hence, irrigation efficiency is seriously compromised. The rigidity or uncertainty of water supplies in many irrigated areas reflects the limitations of inherited irrigation system design concepts and associated technology; for example, most of the surface irrigation projects designed for rice cropping now need to be remodelled for mixed cropping in the face of an overproduction of rice. The challenge here is to develop solutions to the problems of water management offering greater flexibility for adjustments to new cropping systems, which are emerging and are based on the technology of 'The Green Revolution', including new, short-season cultivars, higher cropping intensity, increased use of farm inputs and mechanisation. Increasing cropping intensity and farming flexibility demand total reliability and timely delivery of water to the farm, higher canal capacities, and efficient control on water to meet the higher peak irrigation demands of emerging crop diversification. In this respect, improvement of irrigation technology should be achieved at three levels: the water storage, the conveyance system and the farmers' fields.

The problems at the storage level are being better assessed through the application of modern engineering technology, including satellite imagery and computer models. But this technology has not been used by many of the developing countries. Also, the problems at the conveyance level have been aggravated because efficient technology has not been adopted by many countries. Many conveyance systems suffer because the use of open and unlined canals too often entails large losses of water, mainly by seepage, causes waterlogging of adjacent land and exacerbates the drainage problem. The conveyance systems in many countries, designed originally in accordance with rigid schedules, were well adapted to conditions when cropping intensities were low (about 25 - 50%), when silt contents were high at the points of diversion and when water was generally plentiful. Advances in irrigation design, where storage dams have reduced the silt content of the water at the diversion points and have allowed for increased cropping intensities, require that new investments be made to bring about more flexibility in the conveyance systems in order to supply water on demand and only when the crops need it. Recent development in advancing closed conduits to carry water to the field (under sufficient pressure) may prove to be adequate. However, the capital costs of closed conveyance systems and energy costs of pressurising water, are considered prohibitive in many situations. The remarkable progress made in Israel, in this respect, is worth reviewing in order to assess the cost and benefit of such systems for modernising irrigation in similar agroclimatic zones elsewhere.

The problems of water distribution at the field level are complex. It is here, unfortunately, that we still witness the greatest inefficiency. In this respect, the science of irrigation is making progress in developing methods to monitor the various components of the field water cycle. The efficiency of water management at the farm level cannot be divorced from the efficiency required in the conveyance system. At the farm level, however, other factors interfere, such as social organisations and farmers' participation in the operation and maintenance of irrigation. Many problems affecting irrigation were caused by farmers who were late in building field channels and were negligent in maintaining the system in their regions.

Many public irrigation schemes do not have adequate financial resources to maintain their components, to operate efficiently or to improve water delivery, especially at the field level. The efficiency of irrigation is further reduced by allowing farmers to grow low-yielding crops which could be feasible only because water was made available almost at no cost. This negligence has allowed for the misallocation and wasteful use of water. Only a small fraction of water in many irrigation systems is actually used by the crops. The average efficiency of field application of water in developing countries (measured for rice) is about 32%. The remainder seeps or evaporates from unlined or obstructed canals or over-irrigated fields. Thus many farmers at tail ends of distribution systems suffer from water shortages, especially during the peak-demand period, which lower yields and reduce the benefits of applied farm inputs (mainly fertilisers). In Pakistan, such problems have been addressed at the social levels by establishing and strengthening the water users associations. Greater effort is needed to strengthen this approach and to develop proper training and incentive methodologies designed to increase farmer participation in operation, maintenance and water management. Irrigation engineers have to encourage the involvement of other disciplines in improving the water management system. The problems are not only technically related to hydraulic systems designed and built to operate according to engineering designs - they are operated by farmers and used by social groups which have their own perception of water needs and requirements. This viewpoint has to receive further attention when we design strategies for improving field water management.

Increasing rate of cost recovery

Poor water management has exacerbated the economic performance of many irrigation projects. Insufficient allocation of financial resources for operation and maintenance has aggrevated the problems of water management. A recent World Bank report has emphasised the relationship between the adequacy of operation and maintenance and the willingness of farmers to pay for the service of water distribution. The complex subject of cost recovery has attracted the attention of economists, politicians, administrators and engineers. The main reason given to explain why irrigation projects do not successfully recover the cost of operation and maintenance are: lack of public committment to cost recovery, unreliable water supply due to poor operation and maintenance of irrigation systems, and the often heavy burden of direct and indirect taxes already imposed by governments on the farming sector. Many irrigation projects are in areas where the people are poor, and the issue of the ability of these farmers to pay cost recovery charges constitutes a sociopolitical dilemma for the goernments. However, for the most part, irrigation farmers are better off and have more secure livelihoods than other rural inhabitants and, therefore, an obligation to pay (at least in part) the public sector. Equity criteria are more or less likely to require payment from farmers for irrigation. A second reason for the complexity is that cost recovery is treated differently by different farming communities in different nations. There are different laws, customs and political attitudes towards cost recovery in investment project , particularly irrigation water charges. In some societies, for example, it is a fundamental belief that water should be free, and direct charges for the water per se may not be legal. However, there may not be any prohibition against charging for the services and facilities that are necessary for delivering water to the fields in a timely manner.

It was found, in a recent review of 48 World Bank-financed irrigation projects, that adequate operation and maintenance investment is necessary for a reliable water supply which, in turn, is a necessary but not sufficient precondition for profitable farming. Profitable farming must be a precondition for cost recovery. The review also indicates that efforts to increase water charges are likely to be more successful if they were combined with visible system improvements that result in making the system more reliable and system rehabilitation that can be made conditional upon guarantees of adequate cost recovery.

Several projects have involved establishing farmers organisations (water users associations) to operate and maintain tertiary water distribution systems. In general, only slow progress has been made in establishing these organisations. More could be done in this area to improve the situation by involving the farmers. When designing a cost recovery system, designers should be aware of existing socieconomic and political factors and, in particular, any indirect charges and fares that have already been imposed on farmers. Most important, the cost recovery system should take into consideration the profitability of the farm and the value of commodities being grown under irrigation.

With the present problems in farm income throughout the world, it is unlikely that the beneficiaries would increase their participation in the cost recovery programs. The world is witnessing an enormous increase in the production of almost all agricultural commodities. Indeed, the production of these commodities in the mid-1980s has reached levels so high that the prices have plummeted below actual cost of production. Indonesia, for example, is facing serious problems because of rice surpluses which cannot be exported competitively, given international prices in relation to present production and subsidisation cost and support prices. Thailand is experiencing similar problems with its rice, especially since the American farmers are subsidised to export rice to markets traditionally supplied from Thailand. Malaysia is adjusting its agricultural investment policies because of the decline in the prices of rubber and palm oil. Sudan has not been able to sell much of the 1985 irrigated cotton crop because of saturated markets and low international prices. Farmers producing these commodities cannot sustain themselves without government intervention. They are unlikely to pay the full cost for irrigation.

An approach leading to modernisation of irrigation and increasing crop diversification is receiving increasing attention by the World Bank and concerned countries. The World Bank is addressing this situation through a series of studies to assess investment priorities for increasing flexibility of cropping systems, modernising irrigation technology, and improving the performance of operation and maintenance services. This effort is designed to suit the specific situations and conditions of the countries concerned. For example, while Thailand is supported to modernise its system to increase the efficiency of crop diversification within a system based on rice culture, Sudan will need assistance to increase such efficiency within a diversification effort based on cotton.

IRRIGATION IN AFRICA

The recent drought cycles that affected Sub-Saharan Africa (SSA) in the last two decades have increased interest in the role of irrigation to ensure crop production in drought years. Systematic irrigation in the SSA was not introduced until towards the middle of this century, which is well over a century later than, for example, in India. Currently there are about 5.3 million ha under irrigation in the SSA, of which 2.2 million ha were developed by governments mostly under large-scale schemes (e.g., Gezira in Sudan and Office du Niger in Mali). Commercial companies operate about 0.5 million ha, and the remaining 2.7 million ha are under traditional irrigation. Yields of irrigated crops are usually low, and the use of modern production practices is limited.

The construction of modern irrigation projects in the SSA suffered from poor surveys, inadequate preparation and lack of systematic planning and administration which have led to long construction delays and high cost overruns. Due to weak management and lack of local funds, scheme operation and maintenance have declined, and farmers have not achieved much food security under irrigation. These problems have led to the neglect of irrigated areas, low production and declining yields. Despite these problems, a recent FAO publication concluded that, because of the lack of potential rainfed agriculture and rising demographic pressure on rainfed land, eight of the SSA countries

(Senegal, Burkina, Niger, Mauritania, Mali, Somalia, Botswana and Kenya) - containing about 14% of the SSA population - would most likely make irrigation an essential element of food strategies in the coming years. Before these problems can be solved, farmers will have to organise themselves into water distribution, drainage and reclamation units with a small - but capable, and motivated - government irrigation management service. The absence of such a system has also been a major constraint on crop production.

CROP CHOICES UNDER IRRIGATION IN THE SSA

The main crops produced in the SSA are:under systematic irrigation rice, bananas, sugar cane, cotton and vegetables; under flood irrigation rice; and under receding floods sorghum, mixed food legumes and vegetables. When advancing irrigation in the SSA, one encounters the problem of limited crop choice. The dominant cereal and legume crops are usually grown in the summer season. Wheat and other winter crops are not likely to be successful in most of the SSA countries even where water is available. Improving crop intensity depends on producing short-maturing varieties that can allow farmers to repeat crop cycles once or twice a year. Intensified and improved production of rice and vegetables in several West African countries has been possible, although it has remained on a small scale. Tropical fruit production has also been advanced under irrigation. Somalia, for example, expanded banana production under irrigation in the Lower Shebelli area for export to the Middle East and European markets. This was possible through private initiatives. Production of irrigated groundnuts and cotton has expanded in several SSA countries. Because of the high cost of irrigation development, investment can only be justified if it increases the production of high-value crops. India, for example, uses 73% of its irrigation to produce wheat and rice - both are of higher value than sorghum and millet. In countries experiencing water shortage (California, Australia and Israel), irrigation was developed to produce high-value crops and vegetables for export markets. In these countries irrigation and water application technology is developed mainly for marketable crops grown under intensive cultivation.

Technology for irrigation and water distribution should be sensitive to the needs of crops grown under irrigation. For example, rice (the main irrigated crop in West Africa) does not demand precise irrigation, as do vegetables. Yet, scarcity of suitable land and expensive water urges planners to promote advanced technologies if irrigation projects are are to be economically justified. Other commercial crops that could grow under the SSA agroclimatic conditions, such as sugar, cotton and fruits, require that water application be carefully managed, using modern technology, and a flexible rotation with the traditional crops accepted by the farmers. Rice production has been maintained in several African countries (although at low yields) because of its suitability for traditional irrigation technology which generally does not provide precise control over quantities of water to be applied in response to dynamic crop needs. Such technology can be extremely wasteful of water and may lead to soil degradation by waterlogging and salinisation. The demand for more food, the decline in soil fertility and the limited supply of water urgently require that traditional irrigation and water harvesting systems be modernised, so as to achieve higher levels of sustainable production based on greater economy and water-use efficiency.

New irrigation systems responsive to crop water requirements in the SSA and its marginal climatic conditions have to be developed as we advance our effort to harvest and manage surface and groundwater. Recent development in irrigation technology could make achieving this target possible. The advent of techniques for high-frequency, low-volume application of water in precise response to changing crop needs is spreading in several parts of the world and includes moving systems of sprinkler, drip and micro-spray irrigation, and indeed holds great promise for increasing production efficiency. Properly applied, the new techniques could allow expansion of irrigation into lands once considered sub-marginal(e.g., stony, sandy, sloping or saline).

Problems associated with the above arguments include:

1. New irrigation systems developed in the industrialised countries are too capital-intensive and high-technology-based to be applicable to the low capital, low technology, largely traditionally-run production system of the SSA; hence, the principles of modern irrigation must be adapted and modified, and appropriate technologies must be designed for the different needs of the SSA.
2. Available skills for managing and operating these systems in the SSA are still lacking. The responsibility for establishing and operating these advanced irrigation techniques will have to be borne by the farmers themselves, many of whom have little or no idea of how such systems work.

In addition, limited or meager experience (by government irrigation agencies or dealers) in repairing and maintaining such systems may discourage farmers from investing in the installation of this advanced technolgoy. These issues have received increasing attention as irrigation planners and agronomists realised that large-scale schemes may not be the vehicle for agricultural development in the SSA, and that more effort is needed to advance more efficient small-irrigation systems.

To conclude this section, the scarcity of water in arid regions should justify special efforts to develop water resources (rivers, groundwater and rainfall) in the most economic manner possible. In addition, modern field irrigation techniques currently being advanced in the arid zones of the USA, Israel, Australia and other countries should be promoted in the SSA. They include sprinklers, drip irrigation and the use of porous containers buried in the soil from where water infiltrates towards the root zones. The latter are particularly interesting since they can be manufactured locally. The problem would be the ability of local users to manage and operate these systems, which could be enhanced through proper local training and efficient field supervision. Future investment for improving agriculture in the SSA will have to seriously consider irrigation as a main area for development, including not only modern technology but also development of skilled manpower and capable institutions to operate, maintain and efficiently manage the application of water, the most important but scarce resource in this region.

5 Drip fertigation – a case study of the transfer of technology to traditional Arab vegetable growers

URI OR

INTRODUCTION

Over the past decade, an agricultural revolution has taken place in the Jordan Valley, on the west bank of the River Jordan. The traditional agriculture of the region, that had remained unchanged for hundreds of years and that was based on the cultivation of small plots of vegetables irrigated by flooding, has undergone dramatic transformation in a very short time.

This revolution, that can also be called 'plastics agriculture' has resulted from the introduction of modern systems of cultivation. The major element in this change has been the introduction of drip irrigation installed under plastic mulch. The main crop cultivated is tomato grown off season.

Within a few short years, the yield of vegetables in the region has increased without any increase in the amount of water used or the addition of new areas under cultivation. The data in Table 1 illustrate the changes that have taken place in the production of vegetables in the region during the period 1970 - 1985.

The development of early yielding varieties of watermelon, particularly suited to the region and with a high financial return, has resulted in this crop occupying a prominent place in the cropping pattern with impressive agricultural production.

The Jordan Valley is 20 km wide and 100 km long, extending along the River Jordan, from the Bet She'an Valley, in the north, to Jericho, situated north of the Dead Sea.

Table 1

Production of vegetables, melons and watermelons in the Jordan Valley, 1970 - 1985 (1000 tons)

	1966	*1970*	*1975*	*1980*	*1984*
Vegetables, total yield	48	31	64	105	191
Melons/watermelons only	36[1]	13	24	20[2]	80[2]

[1] Melons and watermelons, rain fed, entire area of Judea and Samaria.
[2] Melons and watermelons under irrigation.

Source: Agricultural Statistics Quarterly, Vol. I-XVI.

The Jordan Valley is the northern section of the Afro-Syrian rift and lies at an elevation of 250 - 396 m below sea level with the Edom mountains to the east and the Judean and Samarian mountains to the west at an elevation of 700 - 900 m above sea level. The River Jordan divides the valley into two regions, the western section which has been under Israeli rule since 1967, and the eastern section which is part of the Kingdom of Jordan.

The valley is a desert area with rainfall averaging from 250 mm in the north to 150 mm in the south. Springs emerging from the surrounding hills provide the water for the region. River Jordan water is not used, since its volume is low and its salinity is high. Most of the river water is trapped in the north by both Israel and Jordan and does not reach the valley. The annual discharge of the western springs is approximately 60×10^6 m^3, which serves for irrigation of the areas near the river and on the hill slopes. The Valley of Jericho is the most well known of these areas. Jericho itself is considered to be one of the most ancient towns in the world. The soil in the valley is heavy and stony, and on the hill slopes it is shallow and stony.

The valley has a desert climate, with temperatures reaching above 40°C in summer and ranging from 10 - 20°C in winter.

The Jordan Valley is part of Judea and Samaria (see maps at the beginning of this book). The majority of the inhabitants are arabs who have been under Israeli rule since 1967.

The main activity in the region is agriculture. The dominant crop is rain-fed olives. Seasonal vegetables for the local market are grown in the western lowland, and in the Jordan Valley. Vegetables both for the local market and export are grown under irrigation and off season.

Table 2 presents data relating to the cultivation of vegetables under irrigation in Judea and Samaria and the Jordan Valley.

Table 2

Area of irrigated vegetables, melons and watermelons in Judea and Samaria and the Jordan Valley (1000 tons)

Year	*Vegetables*		*Melons and watermelons*	
	Total	*Jordan Valley*	*Total*	*Jordan Valley*
1968	3.1	0.8	1.4	-
1972	4.2	1.0	0.3	-
1976	5.2	3.3	0.2	0.02
1980	4.6	2.9	0.3	0.2
1984	5.2	3.8	1.4	1.3

Source: Agricultural Statistics Quarterly, Vol. I-XVI.

The increase in the area of irrigated crops is mainly in the Jordan Valley and the hill slopes. Until the beginning of the 1980s, the crops were mainly vegetables. During the past few years there has been an increase in the cultivation of melons and watermelons on the hill slopes. The vegetables grown are mainly tomatoes, cucumbers, peppers, squash and onions.

The land is owned in part by a number of families living in the towns who lease the land on an annual basis in return for 20% of the value of the yield. A payment of 20% of the value of the yield is also made for water, which generally belongs to the same families. There are also some privately owned farms, mainly in the Valley of Jericho.

Table 3

Distribution of agricultural areas in the Jordan Valley in relation to ownership

District	*Farm type, %*		*Distribution in relation to size (ha)*			
			0.1 - 0.4	*0.5 - 0.9*	*1 - 5*	*above 5*
				%		
Jericho	private	77	72	8	10	10
	leased	23	8	5	74	13
Uja	private	73	3	3	31	63
	leased	27	7	43	36	10
Giftlik	private	27	-	-	-	-
	leased	73	2	68	14	16
Bardala	private	76	18	22	52	8
	leased	24	25	15	55	5

Source: Zemonski, 1977.

Out of a total of 5,100 hectares, approximately 3,000 hectares are leased. Most of the land is owned by some 30 families who also own the sources of water (Regev, 1983).

Irrigation in this region is essential. Rain-fed crops are very few. The water from the springs is carried into the region by canals 20 - 30 km long. These canals were earthen until the 1940s, but since then have been concreted. Distribution channels distribute the water from the central canal to the field. The canal system is owned by all those possessing water rights, with the result that maintenance and improvements are problematic.

There is no correlation between the size of the individual plots and the quantities of water delivered. Water is allocated on a 7 to 8 day roster system. The total discharge of the springs is allocated simulataneously to 1 - 4 users (Agricultural Statistics Quarterly, Vol. I - XVI). The quality of the water varies and some waters are saline. Evaporation rates are high, 17 mm per day in summer, resulting in a high accumulation of salts in the soil.

The bulk of the agricultural produce is marketed out of the region. The dry climate enables the cultivation of vegetables in winter, when crop water requirements are not particularly high. There is a strong motivation to cultivate only in winter when the prices attained are high.

PROBLEMS OF TRADITIONAL FARMING PRACTICES

Until the middle of 1970s irrigation was mainly by flooding. As is the case in many parts of the world, flooding also causes a number of problems in this region (Fukuda, 1976; Merrey, 1982). Notable problems are: low efficiency, salinisation of soil, rise of water table, infestation of soil by diseases, poor distribution of water, soil sealing, soil erosion and weeds.

Of a total of 60 x 10^6 m^3 of water only 40 million reach the fields because of lack of storage facilities and water losses from seepage and evaporation along the canals. Irrigation efficiency in the field is less than 30%. Only on level fields can water be applied by flooding.

The canal system is unreliable. Fields at the tail end do not receive sufficient water so that plots are abandoned, as occurs in other parts of the world with similar systems. In addition, stealing of water and corruption are common (Tiffen, 1984). The allocation of water is rigid and changes can not be made. The initiation of changes in land and water ownership would give rise to severe social upheaval in any place in the world and this is particularly so in this region influenced as it is by the Arab - Israeli conflict (Oakley, 1978).

The irrigation system and lack of any crop rotation resulted in decreased yield per unit area. The average annual yield of vegetables in the 1970s was 10 - 15 ton/ha (Agricultural Statistics Quarterly, Vol. I - XVI).

OBJECTIVES

The accepted practice in the developing countries of introducing changes gradually does not produce the anticipated results (Gittinger, 1982; Framji, 1983; Kassapu, 1979). The introduction of high-yielding varieties such as HYV, relieved the problems of food shortage, but the full potential of these varieties was not fully realised. Their introduction, on the whole, was not accompanied by additional changes in irrigation and fertilisation (Pearse, 1977). In many parts of the world, agricultural production has decreased as a result of the insufficient introduction of new technologies, especially irrigation, or the incorrect use of such technologies. Thus, Punjab in Pakistan, that exported food in the past, has to import food today (Merrey, 1982). The incorrect use of surface systems of irrigation resulted in salinity and the flooding of extensive areas that had to be taken out of cultivation. The introduction of high-yielding varieties in the Punjab only temporarily slowed down the trend of decreasing productivity.

There is an approach today that prefers small-scale irrigation projects to tremendous investments in large projects. This approach can be dangerous in that it may perpetuate an undeveloped agriculture (Kellar, 1980; Storer, 1985).

The introduction of change in developing countries requires re-examination. Irrigation developed that includes a combination of traditional land and water allocation together with technological improvements to the irrigation system and crop cultivation practices is one alternative. An additional approach may be that of an overall 'package deal' to the farmers (Mosher, 1969; Gisin, 1981). This proposal can serve as a model for speedy development. Instead of gradual development over a wide front the concept would be for overall complete and accelerated development in focal points and the spreading of these technologies from the focal points to the periphery (Or, 1988; Gisin, 1981).

CHANGES TAKEN PLACE IN THE JORDAN VALLEY DURING THE LAST 15 YEARS

The improvement in agricultural practices in the Jordan Valley commenced at the beginning of the 1970s. This process was a part of the endeavour of the Israeli Government to narrow the gap between the different groups living in the region and to reduce the economic, social and political tensions between Jews and Arabs in the District. The Israeli settlements along the Jordan were in areas not previously cultivated.
Irrigation water was piped from the Sea of Galilee and the results attained were impressive (Duchan, 1983).

Experiments were carried out in the Regional Experimental Station in the Giftlik to determine methods of cultivation and crops suited to the Arab farmers that would produce optimal yields. The results of a series of experiments indicated that drip irrigation on mulched seed beds was the best system. Both the agricultural results and irrigation efficiency were excellent.

The results obtained in this research project by Prof. D. Goldberg and colleagues from the Faculty of Agriculture are summarised below. The irrigation water used in these experiments was saline and emphasises even more strongly the potential of drip irrigation.

Sprinkler irrigation, which is widely used in Israel, was tried in the Jordan Valley, but failed due to accumulation of salts on the leaves.

Table 4

Yield of vegetables (ton/ha), Regional Experimental Station, Giftlik (Or, 1985)

Irrigation system	*Tomatoes*	*Eggplant*	*Squash*	*Cucumbers*	*Onions*
Flooding	15	20	12	5	15
Drip irrigation	60	70	30	24	35

Within a short period of time a program was initiated incorporating the following elements:

1. Seed - the use of hybrid seed suited to the soil and climate of the region;
2. Soil mulching - the use of transparent plastic sheeting to increase the temperature of the upper soil layers;
3. Drip irrigation under the plastic mulching;
4. Solar sterilisation - exploitation of the high radiation in summer to obtain thermal disinfestation;
5. Low tunnels for the cold season;
6. Constant fertilisation through the drip irrigation system.

Regional agricultural advisers participated in the research and development projects and were confident of the success of the overall proposal. The first approach was to the leading farmers in the district. Three farmers agreed to allocate plots for trial and demonstration purposes, and were guaranteed compensation in the event of failure. Nearby farmers followed progress through field days and frequent visits (Or, 1988). The methods that were used and demonstrated were copied and adopted by additional farmers, some of whom at the beginning conducted trials on small plots. Within a few years the majority of the farmers adopted new systems of cultivation.

The following factors contributed to the success of the program (Or, 1985):

1. The demand for out of season vegetables in Israel and the Arab countries, and the consequent high price levels.
2. Credit given for equipment and materials to the small farmers enabled initial purchases and assistance for continued investment. At first, credit was given by the Israeli Government, but because of political problems, a Church organisation called 'Menonite' took over the provision of credit. The representatives of this organisation provided equipment and not monetary credit. In this way, the farmers used the equipment for the purpose that it was provided. Bulk purchasing resulted in better terms from the suppliers.
3. The local extension service was given 'preferential status' and was supported by the Israeli research and extension services. The local advisers continued to receive advice and information, gathering experience both within the region and further afield.
4. The successful involvement of the companies supplying the agricultural inputs, in particular drip irrigation equipment, is noteworthy. These companies participated in the development of equipment and components adjusted to the requirements of the region. Training programs and courses were also organised for the farmers.

The solutions presented to the farmers were simple and uniform. As a result, the methods were adopted speedily. The key lies in the uniformity of the systems and this facilitated the transfer of information (Or, 1987). The following are the main features of the new methods:

1. Operative storage ponds were constructed alongside the canals and at the same elevation. These ponds were filled by gravitation and in accordance with the traditional allocation rota. Black plastic sheeting was installed on the bottom of the ponds to prevent seepage. The operational use of the ponds did not hinder the traditional allocation of water and did not cause any social problems. The farmers were able to irrigate in accordance with crop water requirements with regard to both timing and size of application.
2. Each storage pond was equipped with standard pumping, filtration and fertilisation equipment. Each unit consisted of a pump, with a discharge of 50 m^3/h at a pressure of 25 m, gravel and control filters to ensure the cleanliness of the water and prevention of emitter clogging, and a fertiliser tank for applying fertiliser through the irrigation system. The number of units per pond depended on the size of the pond.
3. Plastic pipes were used to distribute the water from the control head to the fields. Plastic laterals were attached to the distribution lines and were standard throughout the region. Emitter discharge was 4 L/h and emitters were spaced at a distance of 60 cm along the laterals.
4. Seed beds were uniform, 1.4 m wide with a maximum length of 80 paces (50 m).
5. Fertilisation was standard and consisted of the application of a compound 20:20:20 fertiliser, that included all the essential nutrients. This fertiliser is more expensive than others, but its use ensured that the plant received all the necessary nutrients. The farmer thus did not require any special knowledge as to the composition of the fertiliser or specific plant requirements.
6. The plastic used for mulching in the low tunnels was standard.
7. The number of crops and varieties grown was small and facilitated control by the advisers.

The overall solutions that were proposed assisted in the speedy transfer of instructions from the advisers to the farmers and among the farmers, some of whom were illiterate. Thus it was possible to reach all the farmers during every stage of crop growth. The following cultivation practices were applied uniformly by all the farmers (Or, 1985):

1. August - preparation of seed beds, 1.4 m wide. Organic manure, incorporated into a small furrow in the center of the bed, was intended as a nutrient reserve.
2. Drip laterals were laid out immediately after the preparation of the seed bed, one lateral per bed and covered with plastic mulch.
3. September - application of water for 12 h, twice every 48 h. Due to the high temperature, the water under the mulch evaporates and results in thermal disinfestation and sterilisation of soil from disease and weeds. This technique was developed by Prof. Katan of the Hebrew University and produces clean and disease-free fields.
4. October - cucumbers were seeded, one seed by each emitter.
5. Forty-five days after seeding the cucumbers, tomatoes were planted on the other side of the bed by the emitter. The tomatoes received

the same irrigation and fertilisation as the cucumbers.
6. After the cucumbers had been completely harvested, 70 - 90 days, the tomatoes remained and continued to yield; they were maintained until yield and prices dropped.
7. December to January - after the cucumbers had been cleared, low tunnels were added to afford protection from the low temperatures.

The main crop combination is cucumbers - tomatoes, but there are additional combinations, such as melons and squash instead of cucumbers, and peppers and eggplant, and sometimes onions, instead of tomatoes. Care must be taken to avoid growing two crops of the same botanical family.

In recent years there has been a tendency to grow tomatoes in autumn and follow with watermelons in early spring. At this stage, this combination is very worthwhile.

RESULTS

The following results were attained in the Jordan Valley due to the initiation of these changes:

1. The main most evident result is the increase in yield of the various crops grown.

Table 5

Changes in yields of vegetables under different irrigation methods as compared to the Jewish sector in the Jordan Valley (ton/ha) (Regev, 1983)

	1975		*1982*	
Crop	*Traditional*	*New methods*	*New methods*	*Israeli sector*
Tomatoes	20	40	60	60
Cucumbers	5	15	25	30
Eggplant	20	50	70	70
Onions	15	35	35	35
Watermelons	-	-	70-90	-

The yields in the Arab sector reached the same level as those in the Israeli sector.

2. The influence of the increase in yields and the various inputs resulted in changes in the number of work days invested in the various crops and seasonal activities.

Table 6

Distribution of workdays per hectare for tomatoes in accordance with irrigation methods (Zemonsky, 1977)

Work in days	*Yield*	*Flooding* 15 - 20 tons	*Long furrows* 25 - 30 tons	*Drip* 40 - 70 tons
Irrigation		40	50	10
Cultivation		100	150	120
Harvesting		100	170	220
Sorting, packing		40	70	80
Total, work days		280	440	430

There is an increase in the number of work days for harvesting and packing and a decrease in the number of work days for irrigation and cultivation.

3. The changes that took place in systems of irrigation are impressive. It is emphasised that the areas under drip irrigation are those in which all the innovations included in what is termed the 'package deal' were introduced.

Year	*Area irrigated (ha)*	*Flooding (ha)*	*Drip (ha)*	*Drip (%)*
1970	430	425	5	1.1
1975	1,550	1,295	255	16.4
1980	2,290	350	1,940	84.7
1985	4,170	370	3,800	91.1

Vegetables still irrigated by flooding are marginal in importance and include gardens and fields in transitions.

Until 1967, the average area irrigated in the Jordan Valley was 15,000 ha. In 1970 the area decreased to 500 ha as a result of salinisation. The increase in the area in the 1980s was due to efficient use of water and the lengthening of the agricultural season, enabled most of the water from the spring to be exploited.

4. The increase in production of the various crops should also be noted.

Table 7

Agricultural production in the Jordan Valley 1966 - 1984 (1000 tons)
(Agricultural Statistical Quarterly, Vol. I - XVI)

Crop *Year*	*1966*	*1971*	*1975*	*1980*	*1982*	*1984*
Tomatoes	6	4.6	35.5	45.6	71.9	66.9
Cucumbers	4	6.8	15.2	22.2	22.9	25.9
Eggplant	2	7.5	9.0	17.3	23.7	18.2
Melons/watermelons	36[a]	12.7	3.6	19.6	25.9	80.6
Total	48	31.6	63.3	104.7	144.2	191.6

[a] Yield of melons, watermelons from entire area of Judea and Samaria. In 1984, there was a strong increase in melons and watermelons and a drop in all other vegetables.

5. The changes in yield per hectare had a strong influence on the incomes of the farmers. The results of an economic analysis per unit area are detailed in Table 8.

Table 8

An economic analysis for 1 hectare of tomatoes, cucumbers and watermelons in the Jordan Valley in relation to innovations ($)

	Before change	*After change*	
	traditional farming flooding - tomatoes	*Drip/mulch autumn cucumbers + winter tomatoes*	*Drip/mulch autumn tomatoes, early spring melons and watermelons*
INPUTS			
Cultivation	100	250	250
Plants and seeds	300	800	700
Manure and fertiliser	240	550	550
Plant protection, chemicals	120	440	350
Plastics	-	800	800
Water and pumping	-	120	80
Packaging, materials	80	300	120
Interest	80	330	280
Variables, total	820	3,550	3,130
Return investment, drip	-	450	450
Expenditure, total	820	4,000	3,580
INCOME			
Yield, tomatoes (tons)	20	60	30
Yield, watermelons and cucumbers (tons)	-	25	60
Tomatoes, price (per ton)	200	300	200
Cucumbers, watermelons, price (per ton)	-	300	400
Total productivity	4,000	25,500	30,000
Land leasing and water, 40%	1,600	10,200	12,000
Total surplus for farmer	1,580	11,300	14,420
Surplus, work day	5.6	17.4	20.6

6. There are insufficient data for the region to determine the extent of change in the income per family or person. Information exists as to the increase in the number of tractors and serves as an indication of the extent of investments and improvements in technology. The figures quoted relate to the whole of Judea and Samaria, but it may be assumed that for the most part they relate to the Jordan Valley. Agricultural development in the rest of the area is based mainly on rain-fed olive groves. There is also a move from agriculture to building and industry.

Table 9

Number of tractors in Judea and Samaria, 1970 - 1985 (Agricultural Statistics Quarterly, Vol. I - XVI)

1970	*1973*	*1976*	*1979*	*1982*	*1985*
460	870	1,370	1,880	2,610	3,560

7. The increase in purchased inputs can also serve as an indicator of change. It is accepted that in an advanced agriculture the % of inputs of the total production is higher than in traditional agriculture.

Table 10

Annual expenditure on water, plastics and seeds in Judea and Samaria, 1967 - 1981 (Agricultural Statistics Quarterly, Vol. I - XVI)

	1967	*1970*	*1979*	*1981*
Water	33.8	44.1	118.4	130
Plastics and seeds	16.9	29.4	81.7	139

(Expressed in millions of old Shekels, 1981 prices, in all Judea and Samaria).

In the 1970s, purchased inputs were 15% of the value of production. In 1981, inputs were 40% of the value of production, a level similar to that in an advanced agriculture.

8. The number of inhabitants in the region before 1967 was 76,000; some of these lived in refugee camps, mainly in the Jericho district, and crossed over to the Kingdom of Jordan during the war. At the end of 1967, the population that had remained was considerably lower. Growth was made possible by the uniting of families and the movement of people to the valley.

Table 11

Changes in population in the Jordan Valley area, 1967 - 1984 (Agricultural Statistics Quarterly, Vol. I - XVI)

	1967	*1975*	*1980*	*1984*
Population	12,100	19,800	18,500	23,500

Jericho has a population of about 10,000 inhabitants, some of whom are farmers.

9. Additional changes that can be found in the area are the commencement of building, improvements in the health and educational services, changeover of some of the canals to pipe systems and the use of electricity and radio. Inhabitants that were previously nomadic have settled in villages and built homes.

10. The Bank of Israel periodically reviews the economic development in Judea and Samaria and developments in the Jordan Valley. Following are a number of their findings (Annual report, 1967 - 1984) :

- Growth in agricultural produce during the 1968 - 1973 period was 12% per year. The cultivation of vegetables under irrigation and with plastic protection is increasing. This growth is at the expense of rain-fed melons/watermelons. In 1966, the production of watermelons was triple local requirement.
- In 1970 the import of watermelons from Israel increased tremendously and exceeded local production.
- 1971 Showed an additional decrease in the production of watermelons and an increase in that of vegetables, in particular onions, 43%.
- In 1972 there was a rise of 40% in the price of vegetables grown out of season and this accelerated the growth of vegetables such as tomatoes, cucumbers, eggplant and squash.
- 1974 to 1975 showed a growth in the use of modern agrotechnical methods, such as seeds, irrigation and mulching. There would appear to be an advantage in the growth of winter crops with a high labour requirement.
- 1976 to 1977 showed a continuation of the abandoning of marginal land and switchover to irrigated agriculture, efficient use of water, use of fertiliser and adoption of new varieties.
- 1977 to 1978 showed a growth in agricultural productivity, mainly vegetables. The produce was directed to Israel and to export, particularly of those crops with high labour requirement. The use of modern agrotechnical practices increased.
- In 1981 to 1982 vegetable production increased by 7%. The export of vegetables to Jordan tripled.
- In 1983 to 1984 there was a drop of 2% in production due to limitations of exports to the Kingdom of Jordan owing to a similar development of drip irrigation and mulching of vegetables in the eastern Jordan Valley.
- 1984 noted the beginning of an increase in the production of watermelons. There are no limitations on exports to Jordan.
- In 1985 there was a continuation in the trend of increase of production of melons and watermelons and their export to Jordan.

DISCUSSION

An examination of the agricultural results achieved in the Jordan Valley during the last 15 years demonstrate that development was due to the combined introduction of new cultivation technologies as a 'package deal'. The main technologies were:

1. Drip irrigation;
2. Mulching;
3. Double cropping per season.
4. High yielding varieties.
5. Use of fertiliser.
6. Soil sterilisation.

There were attempts to partially introduce new technologies. The use of local seeds in some fields did not achieve good results. The potential of these seeds was low and they were sensitive to disease.

There still remain small plots that are flooded. These plots are inferior to those irrigated by drip irrigation. Plastic mulching enables earlier yields and solar sterilisation prevents growth of weeds and soil diseases. The lack of the use of any of these practices is a limiting factor and subsequent results are very much lower than in those fields where a complete change was initiated. The farmers themselves preferred the adoption of a total change. The vast majority accepted and used these practices voluntarily, copying the leading farmers.

The discussion in international institutions on gradual development versus immediate and complete change acquires a different significance when viewed against development in the Jordan Valley (Framji, 1983; Gittinger, 1982). Farmers, when given the opportunity will prefer complete change. In the words of one of the farmers 'I do not want to make a lot of changes, one is enough for me'. There are many reasons for the success that was achieved, and relevance to the developing countries requires investigation.

The main reasons for the success were:

1. Demand for produce and good price level;
2. Support system - Israel research and extension services;
3. Existing marketing infrastructure;
4. An efficient credit system that worked without bureaucracy and corruption;
5. A framework for the supply of inputs and services.

Success was attained despite the following problems specific to the Jordan Valley and general problems applicable to developing areas (Oakley, 1978):

1. Limited cooperation with the authorities. The administration was viewed as enforced rule;
2. Lack of desire to introduce any change in allocation of land;
3. Lack of flexibility in systems of water allocation.

It would appear that the attempt to avoid social upheavals or to harm the delicate social fabric was essential. No attempts were made to change the methods of allocation of either land or water. The whole approach was aimed at increasing the productivity and efficiency in the field. The profits made by the owners of the land and water grew, and the questions of social justice were not resolved. However, the increase in income of the farmers led to an increase in the standard of living and to further investment. There may be situations in developing countries where the infrastructure and conditions for accelerated development are lacking and the main test would be the free transition from subsistence farming to an agriculture marketing its own produce. There are, however, situations, mainly around the large cities in the developing countries, where there is a growing demand by the urban population for agricultural produce. In such cases, it would be possible to introduce a similar program of overall change.

In general the initiation of changes in developing countries is effected in stages over a broad front including large populations. Gradual development does not fully exploit the existing potential together with combinations of technologies. The introduction of high yielding varieties (Pearse, 1977) was not a complete success due to the fact that their potential was especially high under irrigation. When irrigation was applied, it became apparent that nutrient deficiency became a limiting factor and consequently it was necessary to introduce the use of fertiliser, as with other methods. Many farmers still grow one crop per year (Hecq, 1981) despite the shorter periods of growth of new varieties. New methods of cultivation will allow for two crops per year and increase productivity and income.

The cumulative experience gained in the Jordan Valley demonstrates the possibilities inherent in a concentrated effort of an overall 'package deal', activated from a focal point and gradually spreading to the periphery and including all the innovations proposed and adopted by the farmer, the successes of one group being copied by others.

Among those concerned with the introduction of new technologies to developing countries, there is a feeling of restraint in the use of systems involving a high level of technology (Bergmann, 1984). The use of traditional technologies perpetuates lack of development. New methods are not only better but are also more reliable. The level of technologies invested in the development and construction of modern methods results in their lasting longer. The greater the effort put into the development of these systems the simpler their operation will be.

At the turn of the century, only a few were able to drive a car. A high level of skill was demanded from the driver. Today, driving is simple, cars are reliable and anyone can drive. The development of an infrastructure of roads and standard road signs make driving easier. The 'road signs' of irrigation, and other practices, must be learnt. Farmers will be unable to adopt new methods without the assistance of a research and extension service. Under existing conditions, the results that were obtained in the Jordan Valley were optimal. It is possible that other conditions would have produced better results. The division of fields into small plots does not allow for any flexibility in cropping practices. Cultivation of plots of 5 - 10 ha would be more efficient. Problems of earthern canals still exist. The optimum solution would be the transport of water open canals to a specific point and from there in pipes with pump and filtration units serving an area of 20 - 40 ha.

Organisation of the use of water and other inputs and marketing arrangements, would improve results, particularly in relation to efficiency and achieving better prices and credit. It may be assumed that in the course of time changes in this direction will take place in the region. Indications of this can already be seen - some canals have already been replaced by pipes and a number of partnerships have been established among families and tribes relating to labour and marketing of the produce.

CONCLUSIONS

The changes that have taken place in the agricultural sector of the Jordan Valley during the past 15 years have destroyed the 'magic circle' of under-development (Pohoryles, 1977). The income per person has risen

From $ 116 in 1966 to $ 660 in 1974 and to a peak of $ 1,000 in 1980. During the last few years, there has been a slowdown in income as a result of a worsening of marketing conditions, the Iran - Iraq war, the Lebanese war and the limitation on imports to the Kingdom of Jordan. The income of the farmers enabled capital investment and additional development of the means of production to be made. The accelerated change at the onset, 16% per year, was due to the initial low level. Later changes were slower.

The changes were impressive. During the first years of concentrated development - real growth of 60% in fertiliser, 650% in the value of spray materials, and 900% in the number of tractors in the region. As already stated, there was a significant change in commencement of building, and medical and educational institutions.

Flooding systems of irrigation that were the exclusive method in 1966 have almost disappeared. In 1975, 15% of the area in the Jordan Valley was irrigated by drip irrigation, today 95 - 98% of the fields are irrigated by drip.

The experience gained during this period in the development of agriculture proves that this information can be used to assist developing countries possessing similar conditions, especially where a developed market opens its doors to a traditional and under-developed market. Similar conditions may be found in various regions of the world where an accelerated urban development has created a demand for food from neighbouring agricultural areas. A slow introduction of technologies into agricultural areas interferes with the orderly development of the urban area and will create a one-way flow of people from the village to the town with the subsequent creation of slum quarters around the towns.

The solutions that have been presented are designed to result in speedy and complete development of regions, rather than gradual development, and are applicable near towns using new technologies such as drip irrigation and the accompanying systems that have been adopted by the Arab farmers in the Jordan Valley.

REFERENCES

Bergman, H., (1984), 'Management structures in Mediterranean irrigation', in *Irrigation Management Network Paper 9C, Agricultural Administration Unit*, ODI, London, 9 - 11.

Duchan, I., (1983), *Jordan Valley Regional Council Development Plan*, World Zionist Organization, Settlement Division, Haifa, Israel, 95 (in Hebrew).

Framji, K.K., (1983), 'Water for food and fibre production: Irrigation - the most consuming water use', *Bull. ICID*, 32(2): 1 - 4, 16.

Fukuda, H., (1976), *Irrigation in the World: Comparative Developments*, University of Tokyo Press, Japan, 329.

Gisin, A., (1981), 'Irrigation in developing countries', *Water and Irrigation Review*, 1(4): 8 - 10.

Gittinger, J.P., (1982), *Economic Analysis of Agricultural Projects*, The Johns Hopkins University Press, Baltimore and London (Second edition, completely revised and expanded, published for the Economic Development Institute of the World Bank).

Hecq, J., (1982), *Problems Connected with Irrigation in Developing Countries*, Conference paper EEC - Israel Seminar on Irrigation, Shefayim, Israel, Commission of European Communities, Brussels.

Kassapu, S., (1979), 'Technology that won't take root', *Cares Fao Review on Agriculture and Development*, 12(2): 41 - 46.

Kellar, J., (1980), *On-Farm Irrigation Practices*, *Advances in Irrigation Technology*, *Proceedings of the World Bank Symposium on Promoting Increased Food Production in the 1980's*, Washington D.C., U.S.A., 291 - 327.

Mayson, E., (1984), *Assisting Villagers to Build their Own Irrigation Scheme - An Example from Thailand*, Irrigation Management Network Paper 9E, Agricultural Administration Unit, Auckland, New Zealand, 13 13 p.

Merry, D.J., (1982), *Reorganizing Irrigation: Local Level Management in the Punjab (Pakistan)*, in: *Desertification and Development: Dryland Ecology in Social Perspective*, Dept. of Anthropology, University of Pennsylvania, Philadelphia, PA, 83 - 109, 363 - 379.

Mosher, A.T., (1969), *Creating a Progressive Rural Structure*, Agricultural Development Council, Inc., New York.

Oakley, P., (1978), 'The Jordan Valley: Drip Irrigation for Small Farming', *Bulletin No. 5*, Reading Development Center, 6 - 10.

Or, U., (1988), *Can Drip Irrigation Development in the Arab Sector of the Jordan Valley Be Used as a Model for Developing Countries*, Hebrew University, Jerusalem, Israel.

Or, U., (1985), E-8 *Jordan Valley Drip Irrigation Scheme - A Model for Developing Countries*, in *Drip/Trickle Irrigation Action*, *Proceeding of the Third International Drip/Trickle Irrigation Congress*, ASAE, 166 - 176.

Pearse, A., (1977), 'Technology and peasant production reflections on a global study', *Development and Change*, 8(2).

Pohoryles, S., (1977), *Agricultural Development in Judea and Samaria Acceleration of Economic Growth. Settlement Geography in Judea and Samaria*, Ministry of Defence, Tel Aviv, Israel, 136 - 145.

Regev, N., (1983), *Arab Section in the Jordan Valley 1975 - 1980*, (in Hebrew), World Zionist Organization, Settlement Division, Haifa, Israel, 27 p.

Storer, J.R., (1985), 'Small scale irrigation - a part solution to the relief of famine in arid areas', *World Crops and Livestock*, 37(4): 126 - 128.

Tibor, G.J., (1983), 'Management versus technology in the development process', *ITCC Review*, 12(3 - 4), 4 p.

Tiffen, M., (1984), 'Human resources in African irrigation, in *African Regional Symposium on Small Holder Irrigation*, Blackie, M.J., (ed.), University of Zimbabwe, 57 - 73.

Zemonsky, E., (1977), *The Arab Farmers in the Jordan Valley, Settlements and Agriculture*, (in Hebrew), Settlement Geography in Judea and Samaria, Ministry of Defence, Tel Aviv, Israel, 605 - 624.

Agricultural Statistics Quarterly, (1967 - 1985), Vol. I - XVI, Central Bureau of Statistics, Jerusalem, Israel.

Bank of Israel, 1967 - 1984 Annual Reports, *The Economical Development of Judaea and Samaria*, Research Department, Jerusalem, Israel.

PART II
MARKETING ISSUES IN ORIENTATION OF OPTIMISING YIELDS

6 Empirical and theoretical aspects in the production and marketing of quality differentiated goods: the case of Israeli tomato export

ODED HOCHMAN AND URI REGEV

INTRODUCTION

This study deals with the internal policy of a governmental marketing agency for an agricultural export product with which many growers are associated. The agency is interested in maximising the expected profits of the whole industry, and operates in a competitive foreign market with a randomly fluctuating hedonic price function. The analysis is, however, immediately applicable to the problem of a firm, facing a given quality price, function subject to uncertainty, that works with profit maximising subdivisions which respond to a management preset price function. The firm must determine the price, as a function of quality, it is willing to pay for the intermediate products (inputs) it uses in the production process. Such a price-quality function is determined on the basis of the contribution of the input to the price of the final product (through its effect on the quality of the latter). Examples of this problem are abundant in sophisticated products composed of many components, such as cars, planes, electronics, fashion goods, wines and most foods, where processing and packaging affect the quality characteristics of the product, in much the same way as inputs in a production process.

The literature on quality-price relationship, started with Hotelling (1929), who developed a competitive equilibrium model dealing with consumers' distribution of preferences for quality. This model was recently investigated by Lancaster (1979). Our paper adopts Griliches' (1971) and Rosen's (1974) hedonic price method to identify the contribution of individual quality components to the price of a product. The method is applied to the problem of pricing tomatoes, whose quality can be measured, as for many other goods, by a binary rule.

A quality trait is of binary nature if each unit of the product either possesses, or does not possess a given level of the quality trait in question. For example, a single tomato is large or not large, firm or not. The quality trait in this case, where many units of the goods are bought in a single transaction, is measured by the percentage of units with that trait, and the overall quality level is a vector of percentages, each for a given trait. The quality trait can also be measured by the probability of obtaining, in a random draw, a single unit of the product possessing the quality trait in question. It is shown below that when quality traits are binary, the competitive market price as a function of quality, must be a convex function, while this cannot be said of a hedonic price function in general.

The convexity of the price function has an important marketing policy implication: a marketing agency facing such a price function would increase its revenue by increasing product differentiation, e.g., the revenue from a given amount of 50% firm fruit would increase if the fruit was sorted into two transactions, one with 25%, and the other with 75% firm fruit. Binary quality traits can be found in many products where the output is subject to quality control with a passing grade threshold; examples can be found in electronics, cars, military equipment, and in agricultural products, one of which is the subject of this paper.

Rapid deterioration in the quality of Israeli export glasshouse tomatoes, followed by price and profit decline, threatened the existence of this industry. A major reason for the decline in quality was considered to be the practice of paying a pooling price to all growers, independent of the quality of their fruit. The industry consists of many small growers, who market their tomatoes via a single marketing agency (Agrexco). The marketing agency sells the product in the Europen market in a competitive environment, under a quantity independent price function, subject to random fluctuations. The only incentive for the growers to produce high quality has been quality control, that disqualified low quality fruit for export.

There exists a short-term (in addition to a long-term) tradeoff in production between quality and quantity; for example, increasing irrigation and nitrogen fertilisation raises the yield while reducing quality. As a result of the pooling price system, together with low bounds on quality, the growers have been led to produce the lowest export quality, thereby reducing the overall profits of the industry. The problem of the marketing agency is to find a solution that will pass on the correct incentives to the individual growers, without diverting from the marketing agency practices of selling one or two brands only, where individual growers are unidentified in the market. The major issues investigated in this study are:

1. Characterisation of the price quality relationship.
2. A pricing policy that will induce optimal production decisions concerning quantity-quality tradeoffs in the production process given deterministic and stochastic market hedonic price functions.

3. Asymmetric information regarding tomatoes quality. The quality control of the tomatoes which is not fully possible on harvest date, but only a few days later. The quality of the produce depends on the growing technique, and is therefore fully known to the grower but not to the marketing agency, which cannot fully determine on harvest day the allocation of the produce to quality classes.

The main results are:

1. We show that in competitive markets and when quality attributes are characterised by binary nature, the Binary Quality Price Function (BQPF) is convex. Implied potential rewards for quality differentiation are explored.
2. A binary quality price function is estimated for glasshouse tomatoes, with significant coefficient of quality variables and high coefficient of determination (R^2 = 0.95). Problems of autocorrelation with missing information were resolved by using an appropriate econometric procedure (Hochman and Regev, 1986). As predicted by the model, the estimated hedonic price function turned out to be convex in the quality traits. The average daily price of tomatoes from the Canary Islands, which was taken as a proxy to the random fluctuation of the market prices, proved to be highly significant.
3. It is shown that the optimal policy of the marketing agency (in a deterministic price situation) is to expose the growers to the market price, or devise a quality premium system, which rewards the growers for quality of their produce in the same way that the market does. Under conditions of price uncertainty, the marketing agency should absorb the risk, and present the growers with a deterministic quality price function equal to the expected value of the market price function. The behaviour of the growers will then match the decision rules implied by the maximisation of the expected profits, this being the goal of the marketing agency.
4. Differentiation of the tomatoes into quality classes by means of a self-selection process is proposed as a solution to the problem of asymmetric information regarding the tomatoes quality on harvest day. This is done by devising a pricing mechanism that will induce the grower to choose the quality class which corresponds to the true quality of his fruit.

The first section of the chapter characterises the price-quality relationship for binary quality, the second section develops the theoretical basis for internal pricing policy by a marketing agency, and the third section presents the empirical estimation of the binary quality price function for glasshouse export tomatoes.

CHARACTERISATION OF THE BINARY QUALITY PRICE FUNCTION (BQPF)

Consider a product, many units of which are bought and sold in a single transaction and whose quality traits are of binary type, that is, each unit either attains or does not attain a given level for that trait. Consequently, a quality trait of the goods can be measured, for any given transaction, by the percentage of units having the desired quality level. The same applies to a vector of quality traits for each of its components. Examples include many food products, clothing items and many other commodities. Clearly, tomatoes are included in this class. For each quality trait, a single unit of the product will take binary values, 1 for possessing the trait and 0 otherwise. The level of each quality trait per unit, say box or kg, will thus be measured by the percentage of tomatoes having that trait. The price paid by the consumers reflects the overall quality of the product as perceived by them; however, the rewards for the individual quality traits are only implicit and cannot be directly observed in the market. The hedonic price method (Griliches, 1971; Rosen, 1974; Brown and Rosen, 1982; Brown and Mendelson, 1984) is applied here to estimate the hedonic price for the quality traits of tomatoes.

Let $Z_i (i = 1,...,n)$ denote the level of the i-th quality trait. For the case of tomatoes this can refer to the percentage of such things as: firm tomatoes, discoloured tomatoes, tomatoes of a given size, tomatoes of a given colour, etc. Then, following Rosen (1974), the competitive market price can be written as a function of the quality traits $P(Z) = P(Z_1, Z_2, ..., Z_n)$, with its partial derivatives being the implicit prices of the quality traits. Estimates of the parameters of this price function can serve as the basic price data for estimating the demand and supply of the quality traits. Furthermore, $P(Z)$ itself is required for the determination of quality premium payments to growers by the marketing agency, based on the quality tests in the packing house. Actually, under deterministic conditions, $P(Z)$ should be the price (premium) received by a grower whose fruit quality turned out to be Z in the quality tests taken in the packing house.

In dealing with the price function for goods with binary quality (BQPF), the following should be noted: (1) full divisibility of the product, and (2) the possibility of mixing and sorting the product. In particular, mixing of given quantities of the product, each with a different quality, will result in a product with a quality level equal to the weighted average of the original qualities. Furthermore, while mixing fruit of different qualities is costless, sorting the fruit into higher and lower quality fruit is costly. This leads to an important property, described in the following proposition:

Proposition: Under conditions of competitive markets, the Binary Quality Price Function (BQPF) is a convex function in its quality traits.

Proof: Competitive markets imply long-run zero profits from arbitrage. We first prove that the BQPF cannot be strictly concave in this situation. To do that, we assume as a working hypothesis, that the BQPF is strictly concave and prove a contradiction. Let A and B be two points on the BQPF (see Figure 1), where:

$$A = (Z^A, P_A),\ B = (Z^B, P_B),$$

$$Z^A = (Z^0{}_1, Z^0{}_2, ..., Z^0{}_{i-1}, Z^A{}_i, Z^0{}_{i+1}, ..., Z^0{}_n),$$

and

$$Z^B = (Z^0{}_1, Z^0{}_2, ..., Z^0{}_{i-1}, Z^B{}_i, Z^0{}_{i+1}, ..., Z^0{}_n).$$

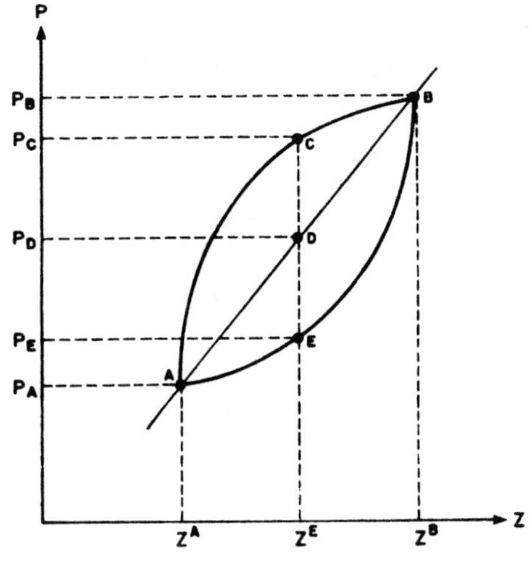

Figure 1 Alternative BQPF curves

Thus, Z^A and Z^B differ from one another only in the i-th element of the quality traits. Suppose that ACB in Figure 1 is the strictly concave market price function, then an arbitrager could buy equal quantities of fruits of qualities Z^A and Z^B, pay P_A and P_B, respectively, and mix them together at no cost. The quality of the resulting product will change only in the i-th element, so that:

$$Z^E = [Z^0{}_1, \ldots, Z^0{}_{i-1}, (Z^A{}_i + Z^B{}_i)/2, Z^0{}_{i+1}, \ldots, Z^0{}_n].$$

Thus the market price of the product will be P_C, while the costs of arbitrage will be $P_D = (P_A + P_B)/2 < P_C$. This implies positive arbitrage gains, in contradiction with the assumption of competitive market, and thus the BQPF cannot be strictly concave.

We next prove that convexity cannot be excluded. Let AEB in Figure 1 be a strictly convex BQPF. If a given quantity with quality Z^E could always be sorted at no cost into qualities Z^A and Z^B, then the hypothesis of strictly convex BQPF (at E) could be rejected under competition. This is so, since then an arbitrager could buy fruit of quality Z^E, sort it into two qualities Z^A and Z^B, sell it for an average price of P_D, and make consistent arbitrage gains, contrary to the competitive market assumption. However, sorting even if possible, is costly (in accordance with the law of entropy) so that this argument fails, resulting in the conclusion that the BQPF can indeed be strictly convex. The difference $P_D - P_E$ is therefore bounded from above by sorting costs.

Furthermore, sorting into two predetermined quality classes, so that the original quality is a convex combination of the two, is often impossible due to interdependence among different quality traits. Suppose, for example, that the i-th quality trait is negatively related to the j-th quality trait. In this case, sorting by the i-th quality trait will change the level of the j-th quality trait as well, in such a way that the 'higher' grade fruit in the *i*-th quality trait will necessarily be of lower quality in the *j*-th component. This implies that in this case Z^E (in Figure 1) cannot be sorted at any cost into Z^A and Z^B. In such a situation the BQPF, and therefore the difference P_D - P_E, will be determined solely by demand and supply conditions, and not by market arbitrage, and may well be as large as P_B - P_A.

The nature of the interdependence among quality traits may differ when different brands are considered; there may even be differences within the same brand among various shipments. Thus in some instances sorting is possible and may be profitable, while in other cases it is not.

In practice sorting is often performed, which by itself implies strict convexity of the BQPF. In addition, quality deteriorates quite rapidly due to the passage of time, handling, weather conditions, etc. Quality changes are not necessarily homogeneous, and sorting into higher and lower grades may be profitable. Indeed we find that sorting for arbitrage gains takes place in different stages of the marketing process. This implies strict convexity of the BQPF in the relevant range, or else sorting would not have been worthwhile (arbitrage gains must be at least as high as sorting costs). This completes the proof of the proposition.

An interesting type of sorting worth mentioning takes place in low-quality low-price retail markets of fresh agricultural produce, especially open markets. Shoppers in these markets put little value on their time, due to low earning power and lack of alternative use of their shopping time. In these markets, we witness cheap sorting done by the customers themselves, who select the better units from low quality produce stands. Clearly, the price charged in such a stand must be higher than that charged for a product of the same quality, where no self-sorting is allowed, since the remaining lower quality merchandise must be sold at a lower price. Such practice is worthwhile only if the BQPF is strictly convex, and the difference between the prices in the two situations is higher than the value of shoppers' time spent on sorting. Convexity of the BQPF has an important implication for marketing policy, as stated in the following corollary.

Corollary : Marketing two quality levels separately will often yield higher profits, and never lower than those obtained by marketing the product at average quality.

This corollary results from the fact that any line connecting two points of a strictly convex function will lie entirely above the function. Thus, the average price of marketing two separate qualities will often be higher and never lower than the price obtained from marketing one brand with average quality. While sorting before packing may be costlier than the additional benefits, mixing different fruit qualities will never be profitable.

INTERNAL PRICING POLICY

The deterministic decision making

In this section we show that the optimal pricing of the produce of the individual producers, who do not directly face the market prices, is to follow the quality-price function. This is the case when a single marketing agency sells the tomatoes of all growers and cannot associate the market prices with individual growers. While the market prices respond to quality differentials, the growers face a pooling price which is the weighted average of all quality prices, without regard to the individual grower's product quality. This practice leads to a free-rider problem: each grower attempts to maximise profits by increasing quantity, regardless of quality. The existence of tradeoffs between quantity and quality leads to a reduction of quality below the level desirable to the growers as a whole. This results in price (and profit) reduction which is external to the individual grower for whom the pooling price is given, independent of the quality of his fruit.

Quality tests at the packing house enable us to determine the quality vector of the produce for each grower. We are looking for a price (quality premium) as a function of quality, $\phi(Z_1, \ldots, Z_n)$, which will be paid by the marketing agency to the grower, according to his produce quality, where each quality component, Z_i, is measured on a scale between 0 and 1. The marketing agency will choose the quality premium (ϕ) that would have yielded a quality-quantity choice by the growers, which would maximise the total net profits of the growers facing the market hedonic price function $P(Z_1, \ldots, Z_n)$.

The formal analysis follows. Denote

$P(Z)$ the binary quality price function (BQPF) where $Z = Z_1, \ldots, Z_n$ is a vector of quality traits.
$F(Z,L)$ the production function - the quantity of tomatoes, with quality Z, produced per unit of land, where L is labour input employed per unit of land.
$S(Z)$ the size of land allocated by the grower to produce tomatoes of quality Z.
N the given number of identical growers.
$\overline{S}$ the size of land per grower.
w wage rate.
$C(Z,S)$ all other costs (beside labour), such as investment in capital stock, land preparation, fertilisers, etc., associated with S units of land allocated to growing produce of quality Z.
$\phi(Z)$ the price quality function set by the marketing agency.

It should be noted that $C(Z,S)$ is independent of labour. This implies the use of a given technology in which $C(Z,S)$ represents the costs of land preparation, while labour is the short-run variable cost.

The problem of the marketing agency is to maximise over $S(Z)$ and $L(Z)$

$$\pi = N \cdot \int_0^1 [P(Z) \cdot S \cdot F(Z,L) - w \cdot L \cdot S - C(Z,S)]\, dZ \qquad (1)$$

subject to

$$\int_0^1 S(Z)dZ - \overline{S} \leq 0. \qquad (2)$$

Equation *(1)* describes the total profits of the N identical growers, and equation *(2)* is the land constraint per grower. It should be noted that the assumption of linear homogeneity in labour and land underlies our definition of $F(Z,L)$ - the quality produced per unit of land.

The conventional assumption of declining marginal productivities is adopted here, as well as the assumption of tradeoffs between quantity and quality, i.e.,

$$F_L > 0,\ F_{LL} < 0,\ F_Z < 0,\ F_{ZZ} < 0 \qquad (3a)$$

where subscripts denote partial derivatives, e.g., the partial derivative $\partial F/\partial L$ is denoted by F_L. We also assume

$$C_S > 0,\ C_{SS} < 0,\ C_Z > 0. \qquad (3b)$$

Note that $C_{SS} < 0$ in *(3b)* implies economies to specialisation in quality.

In practice, special growing techniques are required for each quality, and each technique involves fixed costs in land preparation and capital stock. In addition, the use of different growing methods in close proximity may cause negative spillover effects. These arguments boil down to economies to quality specialisation, which is assumed hence forth.

One way of achieving the goal of maximising π is by direct control of each grower's choice of land *(S)* and labour *(L)* allocated to each quality. Another way is through a decentralised method of setting the quality price function $\phi(Z)$, paid to the individual growers by the marketing agency. The quality of the fruit is further determined in the packing house by quality control tests. The grower allocates land and labour to each quality so as to maximise his profits π_i, where

$$\Pi_i = \int_0^1 [\phi(Z) \cdot S \cdot F(Z,L) - w \cdot L \cdot S - C(Z,S)]dZ \qquad (4)$$

subject to *(2)*.

Obviously, the solution $\phi(Z) = P(Z)$ for all Z will satisfy the marketing agency's goal, since when $\phi(Z) = P(Z)$, the maximisation of net profits by the individual grower, (equation *(4)*), yields the maximisation of the agency's goal in equation *(1)*. Note that $\phi = P$ is only a sufficient condition and not a necessary one.

Given $\phi(Z) = P(Z)$, necessary conditions for attaining the marketing agency's, as well as the grower's goal, are

$$P(Z) \cdot F(Z,L) - C_S(Z,S) - w \cdot L - \lambda \leq 0 \qquad (5)$$

for all Z

$$S(Z) \cdot [P(Z) \cdot F_L(Z,L) - w] = 0 \qquad (6)$$

where λ is the land's shadow price (Lagrange multiplier). It should be noted that while all variables are functions of Z, λ is dependent of Z.

The necessary conditions *(5)* and *(6)* are analysed in Appendix 1, and yield the following:

1. Since $C_{SS} < 0$ (increasing returns to quality specialisation), only a single quality will be selected by the representative grower.
2. The left-hand side of equation *(7)* describes the marginal profits from quality improvement.

$$P_Z \cdot F + P \cdot F_Z - C_Z = 0 \tag{7}$$

$$P_{ZZ} \cdot F + 2P_Z \cdot F_Z + P \cdot F_{ZZ} - C_{ZZ} < 0. \tag{7'}$$

If equation *(7)* holds for some Z_0, and (7') holds as well, then Z_0 is optimal. Corner solution ($Z = 1$ or $Z = 0$) will be optimal if inequality in *(7)* holds for all Z, or *(7')* does not hold. If negative inequality holds in *(7)* for all Z, $Z = 0$ is optimal, while if *(7)* is positive for all Z, $Z = 1$ is optimal.

To conclude this section, we have shown that $\phi(Z) = P(Z)$ is a sufficient condition for achieving the marketing goal, i.e., the price paid by the marketing agency to the growers has to be specified as a function of quality and it should be equal to the market price function. Note that if the growers were indeed identical, they would actually have received a single price. However, this is not a pooling price, since every one of them is aware that if his quality does not reach the desired level, his price will change. When growers are not identical, e.g., they differ in their production or cost function, the result $\phi(Z) = P(Z)$ still holds, yet the growers will obtain differentiated quality levels and prices.

Stochastic variation of the market price

Since tomato is a perishable product with a short shelf life, arbitrage over time and space is quite limited, and each point in space and time constitutes a separate market with a different price. The share of the Israeli tomato in the German market is relatively small, thus their market prices are not affected by their quantities. However, the prices of Israeli tomatoes are affected by the prices of close substitutes, namely other tomatoes which sell in the markets. These price functions are subject to stochastic variations due to changes in their supplied quantities, which depend on weather conditions, or due to stochastic changes in demand parameters.

Let $P(Z,X)$ be the (stochastic hedonic) market price of the Israeli tomatoes, where Z is a quality vector as before, and X is a random variable, which represents this stochastic fluctuation in the market. We selected X to be the ZMP reported price of the leading tomato brand in the market during the growing season - the price of the Canary Islands tomato. As a result, $P(Z,X)$ is a monotone increasing function in X. Let $g(X)$ represent the density function of the Canary Islands tomato price. We can now reformulate the goals of the marketing agency and that of the individual grower.

The marketing agency's problem is to maximise the expected profits [1]

$$\max_{Z,L} \int_0^1 g(x) \cdot [P(Z,X) \cdot F(Z,L) - w \cdot L - C(Z)]dx \qquad (8)$$

This implies that the marketing agency is risk neutral, which is plausible when it is sufficiently large, or a government agency. The necessary conditions for optimum are

$$\overline{P} \cdot F_L = w, \text{ where } \overline{P} = \int_0^\infty g(x) \cdot P(X,Z)dx \qquad (9)$$

$$\overline{P}_Z F + \overline{P} \cdot F_Z = C_Z, \text{ where } \overline{P}_Z = \partial\overline{P}/\partial Z. \qquad (10)$$

Note that these are the same conditions as for the certainty equivalent maximisation problem, which takes fixed prices equal to their expected value, $\overline{P}(Z)$. In order to achieve this goal by a decentralised solution, the marketing agency has to select a price function $\phi(Z,X)$ to pay to the growers. Clearly, setting $\phi(Z,X) = \overline{P}(Z)$ satisfies the marketing agency's goal. This result means that the marketing agency absorbs the uncertainty of the market prices, as it lets the growers face a single deterministic hedonic price function at all times. Thus, the marketing agency also plays the partial role of an insurance company, pooling the risk of many growers. It is interesting to compare these results with those obtained by Leyland (1972) and Sandmo (1971) for conventional production theory that only deals with quantities of homogeneous products. Differences in risk aversion among growers, which will lead to differences in quality, disappear here when risk is eliminated by the marketing agency, and results in more homogeneous quality.

Determination of quality classes by self selection

The results of the former sections imply that it is never profitable to mix batches of different qualities, and revenues are increased by marketing the fruit in two or more quality classes. Thus, quality classes should be determined, and sorting into the different classes should be made if it is sufficiently inexpensive. Some sorting is actually done in the packing house, e.g., into size or colour classes. Sorting by other quality attributes can be done by taking advantage of the fact that quality may differ between growers who make the decision about the quality of their fruit during the growing season by applying various agricultural techniques. In the case of tomatoes, this is done by controlling the irrigation and nitrogen fertilisation levels, where high levels of water and nitrogen result in a high yield, but poor quality.

The problem with this kind of sorting (or avoiding mixing) is that it is difficult to determine the true quality of a batch on harvest day. Quality controllers in the packing house need a delay period of several days to determine the quality of the fruit, while the classification of the fruit has to be made immediately. The grower is the only one who has the full information on his fruits' quality, since he has determined it long before the harvest by his agrotechnical practices. But if asked, the grower has a strong incentive to overstate the quality of his fruit and thus get better prices.

The solution of this quality classification problem with asymmetric information which we propose uses a method of payment which is based both on the declared quality by the grower, and on the actual quality as found from a sample that is examined a few days later. We first show how this price mechanism gives the proper incentives for the case of two quality classes, and then develop a general solution for any number of classes.

The pricing mechanism which induces the grower to reveal his quality for the case of two quality classes is as follows: when the grower brings his produce to the packing house, he has to declare whether his fruit are to be classified as class A or B. The fruit are actually classified and marketed as declared by the grower. At the same time the fruit are sampled and the sample is kept under controlled conditions for several days, and then they are graded on a quality scale of, say, 1 to 10. The price paid to the grower will be a linear function of the quality grade with parameters depending on the declared quality. The function will be steeper for class A, and only moderately sloped for class B (see Figure 2).

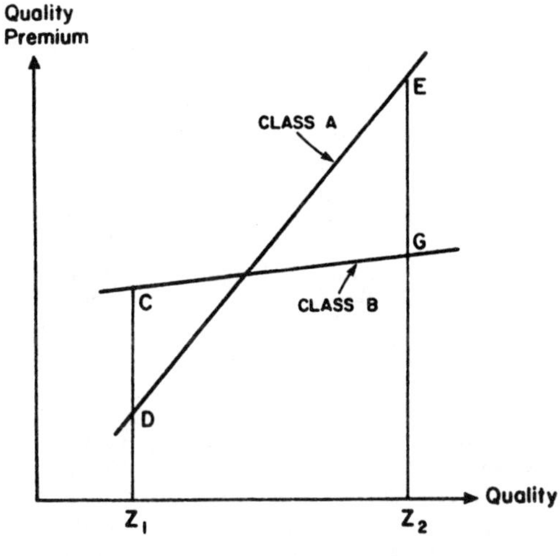

Figure 2 Quality pricing scheme for grading tomatoes into classes by self-selection

This gives greater reward for additional quality grade (or higher penalty for the marginal decline in the grade) for the higher quality class A. Thus, by choosing his quality class, the grower selects the price function he would be paid for his fruit. From Figure 2, it is clear that a grower that knows that his quality grade is Z_1 will never choose his fruit to be classified as grade A, since he will incur a loss of CD in the price of his fruit. A grower with poor quality fruit will therefore choose on his own to classify his fruit as class B, since this choice gives him a higher price. In the same way, if the grower has fruit with quality Z_2, he will clearly prefer to choose class A for his fruit.

This basic idea is extended to any number of quality classes, utilising the knowledge of the quality-price function, and its convexity. Figure 3 illustrates this pricing method.

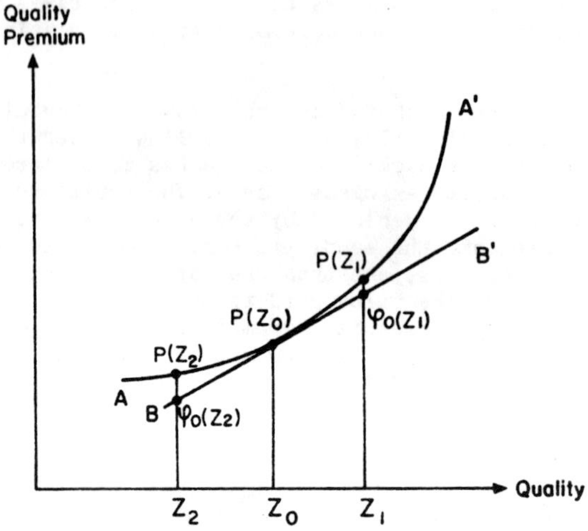

Figure 3 Quality payment function for tomato growers

Let the curve joining A to A' be the price function and let Z_0 be the quality of the fruit that the grower declares. Then this quality determines a price function $\phi_0(Z)$ according to which he will be paid, once the true quality of his fruit is known. The payment function $\phi_0(Z)$ is the tangent plane to the quality-price function $P(Z)$ at Z_0. If the true quality turns out to be Z_0, then the grower gets the price $P(Z_0)$. If, however, the true quality, say Z_2, is lower than the declared one (Z_0), the grower will get $\phi(Z_2)$ rather than the higher price $P(Z_2)$. The latter would be obtained, since by declaring the lower quality Z_2, the grower would have faced a price function $\phi_2(Z)$ which will be tangent to $P(Z)$ at Z_2. The same applies to declaring Z_0 while the true quality is later found to be higher, say Z_1, where the grower receives only $\phi_0(Z_1)$ rather than $P(Z_1)$. The less accurate the grower is in stating his fruits' quality, the higher is the penalty he pays, so that his best policy is to classify his fruit as accurately as he can.

EMPIRICAL ESTIMATION OF A BQPF FOR TOMATO EXPORT

In this section we present the empirical estimation of the BQPF for glasshouse tomatoes exported to West Germany. A distinctive property of the tomato quality price function stems from the fact that tomato quality rapidly deteriorates in time. Sales at different times and locations imply different markets due to the short shelf life of the tomatoes, which makes it prohibitively expensive for arbitrage. This implies that we should observe different BQPFs for different markets (in time and space), which could vary from one another because of random effects. However, such variation could also result in time because of change in supply of close substitutes, or in space due to differences in transportation costs. We assume that these differences in the price-quality functions in the different markets can be represented ad hoc by changes in the maximum price of the most abundant tomatoes in the German markets (in the relevant season), namely the Canary Islands tomatoes. The Canary Islands tomatoes are of relatively stable quality, which is especially true with respect to their top quality fruit which fetches the maximum price. Thus the BQPF of the Israeli tomatoes in the German markets throughout the season will be written as $P(Z,X)$ with $\partial P/\partial X > 0$, where X denotes the price of the Canary Islands tomatoes in the specific market. A simplifying assumption made was that the BQPF is linear in X. The BQPF was allowed to be a third degree polynomial with respect to quality traits, so that strict convexity could be observed. Empirical estimation of the BQPF for Israeli export tomatoes in the German market (March - May, 1979) yields the results presented in Table 1 and Figure 4.

Table 1

Estimation of the Hedonic Price Function

Variable[1]	*Coefficient*	*t-value*[2]
F^2	$.177 \cdot 10^{-3}$	1.3
$F \cdot S$	$-.403 \cdot 10^{-3}$	4.0*
F^3	$.282 \cdot 10^{-6}$	0.3
D	.0484	4.3*
D^3	$-.255 \cdot 10^{-5}$	-2.13*
S	.0155	2.5*
X	2.419	7.9*
$X \cdot F^2$	$-.792 \cdot 10^{-5}$	-.2
$X \cdot F \cdot S$	$-.604 \cdot 10^{-4}$	-1.73
$X \cdot F^2 \cdot S$	$-.137 \cdot 10^{-6}$	-.5
$X \cdot F \cdot D \cdot S$	$.235 \cdot 10^{-5}$	3.35*
$X \cdot F^2 \cdot D$	$-.524 \cdot 10^{-6}$	-1.15
$X \cdot D$	-.04405	-7.16*
$X \cdot D^3$	$.243 \cdot 10^{-5}$	5.78*
$X \cdot D^2 \cdot S$	$-.749 \cdot 10^{-6}$	-2.66*
Intercept	-.0307	-1.91

$R^2 = 0.948$ Durbin-Watson = 2.04

[1] F = percentage of firm tomatoes (average value in the sample = 70.35)
D = 100 minus percentage of discoloured tomatoes (average value in the sample = 86.49)
S = percentage of small tomatoes (average value in the sample = 71.13)
X = price of Canary Islands tomatoes (average value in the sample = 3.13)

[2] An asterisk above a t value means that the coefficient is significantly different from zero at 95% confidence level.

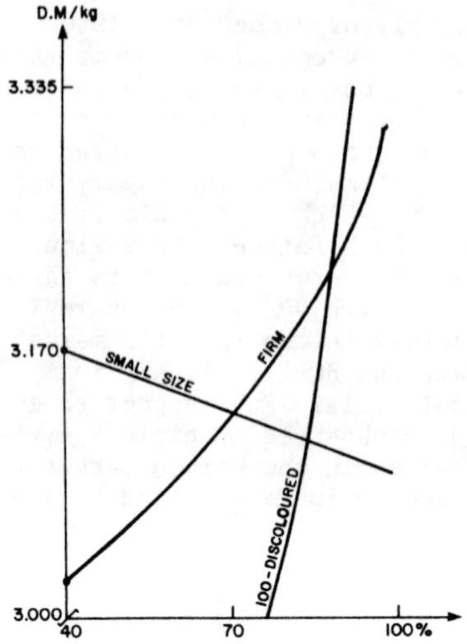

Figure 4 Three projected curve of the BQPF: in each curve, one quality trait is varied and the others maintained at their average values

From Table 1, the matrix of second partial derivatives with respect to quality traits has been evaluated at the average sample values and found to be positive semi-definite, but strictly positive definite when the small size traits *(S)* was omitted. [2] These results corroborate the theoretical arguments concerning the shape of the BQPF. Actually, the degree of convexity is underestimated in our empirical results, since each observation represents an average of a few sales. For example, suppose that an observation is composed of two sales only, then it will lie on a straight line connecting two points on the BQPF, and the function estimated from such points will be closer to linear than the original one. Details of the data, estimation procedures and full results appear in Hochman and Regev (1986).

NOTES

[1] In the previous section, we proved that the grower would allocate all his land to a single quality. Thus, the choice of the optimal quality here replaces the choice of $S(Z)$, which is now trivial, i.e., $S(Z^*) = \bar{S}$; and we assume for simplicity $\bar{S} = 1$, $S(Z) = 0$ for all $Z \neq Z^*$.

[2] From Table 1 and Figure 4, it can be seen that the price function is linear in S (percentage of small size tomatoes). This is apparently inconsistent with the theory which requires strict convexity. However, the tomatoes are sold in the market already sorted with respect to size, so that the prices in our data are actually linear combinations of the prices of transactions of different sizes, where each is homogeneous with respect to size. Therefore, the price function should be linear with respect to size in this case.

APPENDIX 1

Consider the following optimisation problem, which is the problem of both the marketing agency and the growers, after the proper signals have been passed (see text). Maximise over L and S

$$\int_0^1 [P(Z) \cdot S \cdot F(Z,L) - w \cdot S \cdot L - C(S,Z)]\mathrm{d}Z \qquad (A1)$$

such that

$$\int_0^1 S\mathrm{d}Z - \overline{S} \leq 0 \qquad (A2)$$

where the functions and parameters are as defined in the text. The necessary conditions are:

$$S \cdot (P \cdot F_L - w) = 0 \qquad (A3)$$

thus when $S(Z) = 0$, L can be any number.

$$P \cdot F(Z,L) - w \cdot L - C_S(S,Z) - \lambda \leq 0, \qquad (A4)$$

where λ is the non-negative shadow price of the land constraint and is independent of Z. There is inequality in *(A2)* only when $\lambda = 0$. The inequality in *(A4)* is due to an additional constraint requiring $S(Z)$ to be non-negative. Whenever $S(Z)$ is strictly positive equality holds in *(A4)*.

Except for C_S, all terms in *(A4)* are independent of S. Thus $C_{SS} > 0$ means increasing marginal cost of land, which implies diseconomies of specialisation in quality, i.e., if we increase the amount of land allocated to a given Z the marginal cost per unit land increases. Similarly, $C_{SS} < 0$ means decreasing marginal cost of land, implying economies of specialisation in quality.

We first consider the case $C_{SS} > 0$. Assume without loss of generality, that for every Z, $0 \leq Z \leq 1$ where is a strictly positive $L(Z)$, which satisfies *(A3)* with $P \cdot F_L = w$. Also assume that $P \cdot F - W \cdot L - C_S(0,Z) > 0$ for all Z (if it is negative for a given Z, then from *(A4)*, $S(Z) = 0$ for all $\lambda \geq 0$), and let λ be given at some positive value. If for $S = 0$, the left hand side (LHS) of equation *(A4)* exceeds zero (given λ, Z and $L(Z)$), then by increasing S we decrease the LHS, and we do so until we reach an S for which equality holds in *(A4)*. Designate this S by $S(b,Z)$. Let $\tilde{S}(\lambda)$ be $S(\lambda) = \int S(\lambda,Z)\mathrm{d}Z$, and its partial derivative $\partial\tilde{S}(\lambda)/\partial\lambda = \tilde{S}_\lambda \leq 0$. The optimal solution is $[\lambda^*, S^*(Z), L^*(Z)]$ for all Z, where λ^* is such that $\tilde{S}(\lambda^*) = \overline{S}$, $S^*(Z) = S(\lambda^*,Z)$ and $L^*(Z) = L(Z)$.

We now show that in this case ($C_{SS} > 0$), the optimal solution is not unique at a single value of Z and it must include a segment of Z. Assume as a working hypothesis, that the solution (Z_0) is unique, we then show a contradiction. The solution (Z_0) must satisfy equations *(A3)* and *(A4)* with equality when $\lambda = \lambda^*$, and $S(Z_0) = \overline{S}$

$$P \cdot F(Z_0,L(Z_0) - w \cdot L(Z_0) - C_S(\overline{S},Z\ ) - \lambda^* = 0 \qquad (A5)$$

where $L(Z_0)$ satisfies *(A3)*. For all other Z's necessary conditions for optimality imply at least weak inequality in *(A4)* with $S(Z,\lambda^*|Z \neq Z_0) = 0$. By substituting $\overline{S}$ with 0 in *(A5)* we obtain, since $C_{SS} > 0$, the following

$$P \cdot F(Z_0,L(Z_0)) - w \cdot L(Z_0) - C_S(0,Z\ ) - \lambda^* > 0 \qquad (A6)$$

Now, if we replace Z_0 with Z' in *(A6)*, then *(A6)* must remain positive for Z' sufficiently close to Z_0. This contradicts *(A4)*, for $Z' \neq Z_0$, and and therefore the working hypothesis must be rejected, which implies that a whole segment of Z is optimal.

The more realistic case is the one of economies of specialisation in quality, in which $C_{SS} < 0$. In this case, only a single optimal value of Z exists. To show this we assume, as a working hypothesis, that more than one optimal value of Z exists, and prove a contradiction. Let Z_1 and Z_2 be any such pair in the optimum for which $S^*(Z_1) > 0$ and $S^*(Z_2) > 0$. Being in the optimum, both Z_1 and Z_2 satisfy *(A4)* with equality, hence

$$\begin{aligned} &P(Z_1) \cdot F(Z_1,L(Z_1)) - w \cdot L(Z_1) - C_S(S(Z_1),Z_1) = \\ \lambda^* = \ &P(Z_2) \cdot F(Z_2,L(Z_2)) - w \cdot L(Z_2) - C_S(S(Z_2),Z_2). \end{aligned} \qquad (A7)$$

It is easy to see, upon observing the goal function in *(A1)*, that increasing (decreasing) the land allocated to Z_1 by one unit, increases (decreases) the value of the goal function by the amount specified on the LHS of *(A7)* which equals λ^*, and the same is true with respect to Z_2 and the RHS of *(A7)*. If we transfer a unit land from Z_2 to Z_1, then the following inequalities hold

$$C_S(S(Z_1) + 1,\ Z_1) < C_S(S(Z_1),\ Z_1)$$

and

$$C_S(S(Z_2) - 1,\ Z_2) > C_S(S(Z_2),\ Z_2)$$

because $C_{SS} < 0$. Therefore, the equality in *(A7)* will now change into an inequality, and read

$$\begin{aligned} &P(Z_1) \cdot F(Z_1,L(Z_1)) - w \cdot L(Z_1) - C_S(S(Z_1) + 1,Z_1) \\ &> P(Z_2) \cdot F(Z_2,L(Z_2)) - w \cdot L(Z_2) - C_S(S(Z_2) - 1,Z_2). \end{aligned} \qquad (A8)$$

This means that the goal function increases by the LHS of *(A8)* and decreases by the RHS of *(A8)*, i.e., the goal function increases by the difference between the LHS and the RHS of *(A8)*, which is now *positive*. But this contradicts our initial assumption that we are in the optimum, since if we were, we would not have been able to improve the value of the goal function, as shown above. Hence, when $C_{SS} < 0$, it is impossible to have an optimum in which both $S^*(Z_1) > 0$ and $S^*(Z_2) > 0$, when $Z_1 \neq Z_2$. That is, all land is allocated in this case to growing only one quality, Z^*.

Thus we know that $S(Z^*) = \overline{S}$, all we have to determine is the value of Z^*. Hence S ceases to be a variable, and Z becomes a variable instead. The single optimal quality can be determined by differentiating the goal function *(A1)* with respect to Z, with $S = \overline{S}$ (we further assume $\overline{S} = 1$, for the sake of simplicity). The new necessary condition for an internal solution is

$$P_Z \cdot F + P \cdot F_Z - C_Z = 0 \qquad (A9)$$

If, however, $P_{ZZ} \cdot F + P \cdot F_{ZZ} + 2P_Z \cdot F_Z - C_{ZZ} > 0$, or if the LHS of *(A9)* is always different from zero, then we have corner solutions, i.e., $Z^* = 1$ or $Z^* = 0$, depending at which end of the quality segment the goal function attains a higher value.

REFERENCES

Brown, G. and Mendelson, R., (1984), 'The hedonic travel cost method', *Rev. Econ. Stat.*, 66: 427 - 433.

Brown, J.N. and Rosen, H.S., (1982), 'On the estimation of structural hedonic price models', *Econometrica*, 50: 765 - 768.

Griliches, Z., (ed.), (1971), *Price Indexes and Quality Change: Studies in New Methods of Measurement*, Cambridge, Mass., Harvard University Press.

Hochman, O. and Regev, U., (1986), *Characterization of a Hedonic Binary Quality Price Function with Application to Israeli Export Tomatoes*, Monaster working paper, 86-5.

Hotelling, H., (1929), 'Stability in the competition', *Economic Journal*, 39: 41 - 57.

Lancaster, K., (1979), *Variety, Equality and Efficiency*, Columbia University Press, New York.

Leyland, H.E., (1972), 'Theory of the firm facing uncertain demand', *Amer. Econ. Rev.*, 62: 278 - 291.

Rosen, S., (1974), 'Hedonic prices and implicit markets', *J. Pol. Econ.*, 82: 34 - 55.

Sandmo, A., (1971), 'On the theory of the competitive firm under price uncertainty', *Amer. Econ. Rev.*, 61: 65 - 73.

This research was supported by a BARD Grant No. 1-387-81.

7 Consumer and wholesale preferences for fresh vegetables, especially tomatoes, in the Federal Republic of Germany

REIMAR VON ALVENSLEBEN

SOME GENERAL DEMAND TENDENCIES

In a growing economy, variables such as income and prices are losing relative importance as factors determining changes in the demand for food, while other factors such as consumer attitudes and preferences are gaining relative importance. In a high income economy like Germany, future changes in demand will mainly be caused by changes in consumer attitudes and preferences and not so much by changes in income and prices. Although the income elasticities of the demand for vegetables are rather low, for the more expensive items they are above average (see Table 1).

The total demand for vegetables seems to have reached saturation point (Figure 1), however, the assortment is changing (Table 2) and there is a clear tendency towards more diversity in demand. Bulk items such as cabbage, cauliflower and beans are losing relative importance, while onions and especially the 'other vegetables' category containing most of the Mediterranean vegetables (sweet pepper, eggplants etc.) are gaining popularity. Fresh tomatoes exhibit average performance. Further general demand tendencies are summarised in Table 3.

Table 1

Average income elasticities[a] of the household expenditure for vegetables per household size in Germany (FR), 1978

Product / Household size	Number of persons 1	2	3	4	5 and more
Fresh, total	0.30	0.35	0.37	0.42	0.47
cauliflower	-0.15	0.11*	0.16	0.40	0.49
cabbage	-0.21*	-0.05*	-0.11*	0.31	0.50
carrots	-0.15*	0.18	0.32	0.33	0.44
asparagus	0.86	0.75	1.05	1.42	1.49*
lettuce	-0.07	0.15	0.25*	0.24	0.33
tomatoes	0.52	0.35	0.29	0.22	0.55
cucumbers	0.31	0.22	0.19	0.28	0.24
onions	-0.10*	0.10*	0.21*	0.29	0.32
other	0.33	0.43	0.45	0.47	0.46
Preserved, total	0.28	0.31	0.20	0.24	0.28
canned	0.22	0.27	0.18	0.20	0.23
dried	0.47	0.68	0.79	-0.12*	0.45*
frozen	0.68	0.70	0.38	0.52	0.43

* not significant.

a calculated on the basis of a household survey by the Federal Statistical Office.

Source: Kreienhop, S., (1984), *Bestimmungsgründe der Nachfrage nach Obst und Gemüse in privaten Haushalten*, Hannover.

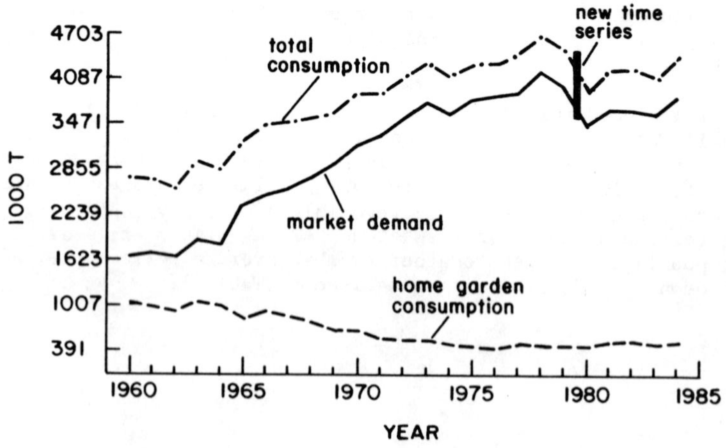

Figure 1 Development of vegetable consumption in Germany (FR)

Table 2

Structure of the vegetable consumption (in %) in households with medium income in Germany (FR)

	Quantity				Value			
	1965	1975	1980	1985	1965	1975	1980	1985
Vegetables, total	100.0	100.0	100.0	100.0	100.0	100.0	100.0	100.0
processed and dried								
frozen	2.6	3.1	3.4	4.9	4.1	4.6	5.1	7.1
fresh	72.8	63.1	62.0	60.6	65.5	57.8	58.8	56.0
Fresh vegetables								
cauliflower	8.4	6.7	5.9	4.8	8.2	6.0	5.4	4.5
cabbage	17.1	12.6	12.2	11.1	10.4	6.7	6.5	5.5
carrots	10.9	9.9	10.2	10.2	7.2	6.2	6.1	6.1
tomatoes	16.1	16.4	15.0	16.8	19.7	20.4	18.7	18.7
cucumbers	7.2	12.2	12.4	11.8	7.4	11.0	12.0	11.4
beans	3.6	2.0	1.5	1.0	3.6	2.2	1.9	1.2
peas	0.4	0.2	0.2	0.1	0.5	0.2	0.2	0.1
spinach	1.7	0.5	0.4	0.2	1.5	0.5	0.3	0.3
lettuce	10.3	10.8	10.5	8.7	15.3	15.8	14.2	13.5
onions	9.7	11.8	13.1	15.0	5.8	6.8	8.3	7.4
other	14.6	16.9	18.5	20.3	20.5	24.2	26.4	31.2

Source: Federal Statistical Office.

Table 3

Demand tendencies for fruit and vegetables in Germany (FR)

	Past	Future
Total demand	Increasing - high income elasticity - increase of population	Stagnating, decreasing - low income elasticity - decrease of population
Diversity of demand	increasing	increasing
Seasonal differences	decreasing	minor changes
Percentage of processed products	increasing	minor changes
Percentage of juices, frozen products	increasing	increasing
Home garden production	decreasing	stagnating, increasing

PURPOSE AND METHOD OF INVESTIGATION

Against the background of these general demand tendencies, the aim of the investigation was to determine to what extent preferences exist for 1) vegetables of different countries of origin and 2) vegetables grown under glass rather than in the open. For this purpose the following surveys were conducted:

1. A representative consumer survey of 400 housewives in Hannover.
2. A sensoring test (visual evaluation and taste trial) involving 90 of the housewives who had already participated in the consumer survey.
3. A survey of 50 fruit and vegetable wholesalers.

The results of these surveys are described in detail by Deters (1985). This paper summarises some selected results.

GENERAL CONSUMER TYPOLOGY ON THE MARKET FOR FRESH VEGETABLES

Consumers differ in their attitudes and behaviour. An average or typical consumer is nonexistent. For planning marketing strategies, it is important to know in which way consumers differ from each other. To supply this information, consumer typologies have been developed, such as the consumer typology of the market for fresh vegetables shown in Table 4.

Table 4

Consumer typology on the market for fresh vegetables in Hannover:
--- below average, +++ above average

Segment	A (14%)	B (23%)	C (21%)	D (17%)	E (14%)	F (11%)
1. *Attitudes*						
- preferences for open field to greenhouse vegetables	---	---	+	++	++	+++
- pronounced demand for external quality	++	Ø	---	++	+++	---
- reservations against canned vegetables	++	-	--	---	+++	+++
- desire for advise and information	++	---	Ø	+++	Ø	Ø
2. *Sociodemographic criteria*						
- age	Ø	--	--	+++	+++	-
3. *Consumption behaviour*						
- share of purchase						
in supermarkets	Ø	+++	++	-	--	---
in weekly markets	Ø	++	-	Ø	++	+++
- consumption intensity of fresh vegetables	Ø	Ø	--	Ø	++	+++

Source: Deters, S. (1985).

With the help of cluster analysis, six market segments have been identified which differ with respect to four attitude vegetables parameters:

1. Preferences for open field over greenhouse produce.
2. Pronounced demand for external quality.
3. Reservations against canned vegetables.
4. Desire for advise and information.

For instance, the F market segment (11% of the sample) exhibits the highest consumption intensity of fresh vegetables. These consumers have the strongest preferences for open field vegetables, have a less pronounced demand for external quality, have strong reservations against canned vegetables and an average desire for advise and information. Many of them prefer to buy in weekly markets rather than in supermarkets.

The E market segment (14% of the sample) has the second highest consumption intensity and similar attitudes and consumption behaviour, however, these customers regard external quality to be of paramount importance. A factor to be noted is that their age is far above the average. It would be very unwise to offer vegetables labeled as 'glasshouse products' to E or F segment customers. In general it is inadvisable to inform consumers about the production method if vegetables have been grown in a glasshouse, since about two-thirds of the consumers have reservations against glasshouse cultivation. This tendency is increasing and even the Chernobyl-accident has not changed these consumer preferences. Approximately 90% of the vegetables in Germany are grown in the open field, and German growers are being advised to label their vegetables accordingly.

PROPERTIES OF AN "IDEAL" TOMATO IN DIFFERENT MARKET SEGMENTS

The usefulness of market segmentation becomes more evident, if we look at the survey results on the tomato market (Table 5). Again, six market segments have been identified differing with respect to the following four evaluation dimensions of the 'ideal' tomato:

I Tomato of high external quality versus tomato of low external quality.
II Glasshouse tomato versus tomato grown in the open field.
III Naturally-produced tomato versus unnaturally produced tomato.
IV Tomato with a Mediterranean character or southern appeal.

For instance, consumer Type 4 (13% of the sample) has the strongest preference for a high external quality and for glasshouse tomatoes and dislikes Mediterranean tomatoes. For these consumers, the Dutch tomato comes closest to the ideal tomato, the Mediterranean tomato is farthest from the ideal and the German product ranks in between. However, this is the only segment with a strong preference for Dutch tomatoes. The salient results are:

1. Two of the six segments (42% of the sample) are not interested in the country of origin.
2. One segment (20%) likes German and dislikes Dutch tomatoes.
3. Two segments (25%) prefer Mediterranean tomatoes and dislike Dutch tomatoes.
4. No segment dislikes the tomatoes of its own country.

Table 5

Consumer types with homogeneous requirements regarding an ideal tomato. Order of preference of tomato from different countries

Consumer types	Evaluation dimensions of an ideal tomato I	II	III	IV	Ranking of preference (Euclidean distances from the ideal tomato)	
Type 1 (29.5%)	---	0	--	-	1. German tomato	1.1
					2. Mediterranean tomato	1.3
					3. Dutch tomato	1.6
Type 2 (12.3%)	-	+++	++	+	1. German tomato	1.0
					2. Dutch tomato	1.2
					3. Mediterranean tomato	1.5
Type 3 (20.3%)	++	---	+++	-	1. German tomato	1.5
					2. Mediterranean tomato	2.1
					3. Dutch tomato	2.6
Type 4 (12.8%)	+++	+++	-	---	1. Dutch tomato	0.4
					2. German tomato	1.2
					3. Mediterranean tomato	2.5
Type 5 (11.6%)	--	--	+	+++	1. Mediterranean tomato	1.4
					2. German tomato	2.0
					3. Dutch tomato	2.2
Type 6 (13.3%)	++	-	-	+++	1. Mediterranean tomato	1.4
					2. German tomato	2.0
					3. Dutch tomato	2.2

Evaluation dimensions:

I Tomato of high external quality (+) vs tomato of low external quality (-).
II Glasshouse tomato (+) vs tomato grown in the open field (-).
III Naturally produced tomato (+) vs unnaturally produced tomato (-).
IV Tomato with a Mediterranean character or 'southern appeal' (+) vs tomato without a Mediterranean character of 'southern appeal' (-).

0 indifferent
\+ positive loading } modified according to intensity
\- negative loading }

Source: Deters, S. (1984).

It is surprising to note that the image of the Dutch tomatoes, which is dominating the German market, is not as good as is generally believed.

For the retail trade, the following conclusion can be drawn: to maximise consumer satisfaction, it is advisable to offer a range of different tomatoes - glasshouse as well as open-field products and tomatoes from various countries.

SENSORY TESTS WITH TOMATO: EFFECT OF LABELS

Some sensory tests were conducted to indicate the possible effect of labeling tomatoes with the country of origin and the production method.

Visual evaluation of different samples yielded the following results (Table 6) :

Table 6

Visual evaluation of tomato samples with and without labels (n=89)
Average scores: 0 = lowest, 5 = highest score

Evaluated tomatoes	*Unlabeled*	*Labeled*	
German open-field	3.1	'Germany'	3.4
German 'organic'	2.2	'Germany' grown organically	3.7
Italian	3.5	'Italy'	3.4
Belgian beef	3.5	'Belgium'	3.3
Dutch	3.5	'Holland'	3.2

Source: Deters, S. (1985).

1. Unlabeled tomatoes of the poorest external quality receive the lowest scores.
2. The label 'Germany - organic production' has a great positive effect on consumer expectation.
3. The label 'Germany' has a small positive effect and the label 'Holland' a small negative effect on the consumer expectation.

The results of taste trials involving the same test persons can be summarised as follows:

1. Many consumers are able to recognise taste differences.
2. A test using identical samples with different labels showed that the label 'Germany-grown in the open field' has a positive effect on the taste trial scores as compared to the labels 'Holland', 'Italy' and 'Germany'.

These results confirm that it is advisable to identify home-grown and open-field tomatoes by appropriate labeling, since such products are preferred by some consumers.

WHOLESALER PREFERENCES FOR TOMATOES

The final step of the investigation was a survey of 49 wholesalers in Germany to determine to what extent the wholesaler preferences coincide with consumer preferences.

Evaluation criteria

The most important aspect determining the buying decision of a wholesaler is the external quality of tomatoes. Taste, price and handling are of secondary importance. Wholesalers were asked to rank the evaluation criteria which are important for their buying decisions. The following average ranks were given:

External quality	1.3
Taste	2.5
Price	2.7
Handling	3.1

The term 'Handling' comprises all aspects of distribution and marketing logistics, such as sufficient quantities, uniform packs, continuity of delivery, possibility for short-term orders, sticking to delivery dates etc. Larger firms rank the 'Handling' aspect higher than do small firms.

External quality

With respect to the external quality of tomatoes, the various supplying countries have been evaluated as follows (0 = lowest score, 5 = highest score):

1.	Holland	4.2	7.	Germany, open-field	3.2
2.	Belgium	4.1	8.	Morocco	2.7
3.	Germany, glasshouse	3.9	9.	Italy	2.4
4.	Canary Islands	3.8	10.	Albany	2.2
5.	Israel	3.7	11.	Romania	2.2
6.	Spain, mainland	3.3	12.	Bulgaria	2.1

Taste

With respect to taste, Dutch tomatoes have an image far below average. Tomatoes from the Canary Islands and open-field products from Germany receive the best scores:

1.	Canary Islands	4.4	7.	Italy	3.3
2.	Germany, open-field	4.1	8.	Morocco	3.3
3.	Belgium	3.9	9.	Albany	3.0
4.	Germany, glasshouse	3.7	10.	Bulgaria	3.0
5.	Spain, mainland	3.5	11.	*Holland*	3.0
6.	Israel	3.4	12.	Rumania	3.0

Handling

As far as the handling aspect is concerned, Holland is far ahead of all other countries (1 = best score, 4 = lowest score):

1.	Holland	1.0	7.	Italy	2.4
2.	Belgium	1.7	8.	Morocco	2.6
3.	Canary Islands	1.8	9.	Albany	2.9
4.	Spain, mainland	1.8	10.	Romania	2.9
5.	Germany	2.1	11.	Bulgaria	3.2
6.	Israel	2.3			

Desired properties of tomatoes

Wholesalers were asked which properties the tomato consumer is looking for. The answers have been ranked according to their frequency (in %) as follows:

Firmness	96	*Origin*	10
Taste	88	*Open-field production*	8
Even colour	80	Not packed	8
Smooth	76	Glasshouse production	2
Even shape	43	No residuals	2
		Prepacked	2

As can be seen from these answers, most of the wholesalers do not believe that the country of origin or the production method are properties that the consumer considers important. However, it was found that:

1. 47% of the wholesalers believe that the origin influence the buying decision of the consumer.
2. 18% agree partially to this opinion.
3. 33% do not believe that the origin has an influence on the consumer.

Wholesalers were asked if they believe that the label 'Open-field tomato' has an effect on the buying decision of the consumers. The following answers were given:

1. 53% of the wholesalers expect a positive effect.
2. 45% expect no effect.
3. 2% expect a negative effect.

Consumer preferences for tomatoes of different origin

Wholesalers were asked their opinion on consumer's evaluation of tomatoes of different origin. The following average scores were given (0 = lowest score, 5 = highest score):

1.	Holland	4.5	7.	Israel	3.5
2.	Belgium	4.0	8.	Morocco	3.2
3.	Canary Islands	4.0	9.	Italy	2.7
4.	Germany, glasshouse	3.9	10.	Albany	2.5
5.	Germany, open-field	3.7	11.	Bulgaria	2.5
6.	Spain, mainland	3.5	12.	Romania	2.5

Despite the poor taste image, the Dutch tomatoes received the highest scores. Many wholesalers do not believe that consumers notice any taste differences between tomato origin. According to the experience of the wholesalers, the external quality is the dominating buying criterion. It is believed that this is the reason that Dutch tomatoes are the most preferred tomatoes on the German market.

Sale prospects

The sales prospects for tomatoes in general, for beef tomatoes and for tomatoes of different countries have been scored as follows (1 = best score, 5 = lowest score):

Tomatoes, in general	2.1
Beef tomatoes	1.6

1.	Canary Islands	2.0	7.	Israel	2.7
2.	Holland	2.0	8.	Morocco	2.8
3.	Belgium	2.1	9.	Albany	3.4
4.	Spain, mainland	2.4	10.	Italy	3.4
5.	Germany, glasshouse	2.5	11.	Bulgaria	3.5
6.	Germany, open-field	2.6	12.	Romania	3.5

The demand for beef tomatoes is assessed to be higher than the demand for tomatoes in general. The countries with the highest scores are Canary Islands, Holland and Belgium - the same countries which are believed to be producing tomatoes which best meet the consumer demand.

CONCLUSION

The following conclusions can be drawn:

1. Consumer preferences and attitudes vary greatly. An average or a typical consumer does not exist. For this reason, it is useful to design marketing-strategies taking account of different market segments. To maximise consumer satisfaction, the retail trade should offer a range of products of different origin and different production methods.
2. Many consumers have preferences for home-grown and open-field produce. For this reason a label, informing the consumer of the country of origin and the production method, has a positive (negative) impact on the buying decision.
3. Wholesaler preferences do not always coincide with consumer preferences. In particular, the consumers' image of Dutch tomatoes is not as good as many wholesalers believe. For wholesalers, the external quality and handling aspect seem to be a much more important buying criterion than taste.

REFERENCES

Deters, S., (1985), *Analyse der Verbraucher- und Händlerpräferenzen bei Frischgemüse. Forschungsberichte zur Okonomie im Gartenbau,* Heft 55, Hannover and Weihenstephan, 309 p.
Deters, S., (1984), 'Consumer preferences for fresh vegetables in Germany (F.R.)', *Acta Horticulturae,* 155:211 - 215.

8 The United Kingdom market for salads

LYNN HINTON

INTRODUCTION

The salad market in Britain comprises tomatoes, cucumbers and lettuce in the main, and celery, asparagus and minor commodities such as aubergines, peppers and Chinese leaves. Tomatoes dominate the group with an annual apparent consumption per head of 7 kg, lettuce is next in importance, having an annual consumption of over 4 kg per head, and cucumbers follow with an annual consumption of nearly 2 kg per head. The three salads are all characterised by large variations in seasonal consumption, the lowest in autumn and winter and the highest in spring and summer, but the distinct seasonal profile is in part determined by supply. Demand for salads is on the increase in Britain.

This chapter considers two main themes: firstly, the demand for salads in Britain, and secondly the competition in supply, taking into account supply from countries within the EEC, dominated by Holland, and that from third countries (those non-EEC countries which have trading status with the EEC), and their chances in this expanding market. Considering this competition, successful timing of supply is important, but quality produce is paramount. These and other relevant points will be developed.

With regard to market shares, the salad market is dominated by tomatoes at 54 per cent of consumer spending, followed by lettuce at 19 per cent, cucumbers at 14 per cent, spring onions (seldom imported) at five per cent, and finally radishes (also only token imports), watercress and beetroot, which together account for five per cent. Celery takes three per cent of consumer spending. There has been a shift in the distribution pattern at the retail end. Multiple growers now have 40 per cent of the sales, an increase of three per cent from three years ago, and this has been lost by traditional greengrocers, who now have 30 per cent. Market stalls remain steady in their share at 17 per cent.

The market remains dominated by the three most important salads: tomato, lettuce and cucumber, and, some way behind, celery. Less popular salads now finding a foot in the market are the exotics, mooli, radicchio and endive, as well as peppers, together with the multicoloured capsicum. From just over seven per cent in 1983, Iceberg lettuce has doubled its share and, with its premium price and extra weight per piece, it now accounts for more than a quarter of all lettuce sales. Overall sales of lettuce have not expanded and Iceberg has progressed at the expense of other crisp varieties, which have dropped their share of the lettuce market from 43 to 37 per cent since 1983. Even though beef and cherry tomatoes are more available now, 98 per cent of sales of tomatoes are still of the standard salad type.

Consideration will be given to supply and demand from 1971 to 1985 to show developments over time and the changes in the home supply and import components. A full analysis with comment is given for the 1985 season for each commodity, dealing with the main sources of imports in detail. Home market price information for 1985 is discussed in relation to the seasonal pattern of supply, in the context of competition which is so basic to this delicate market, where price is associated with quality.

DEMAND FOR SALADS

Serviceable data on the demand for salads comes from the official Annual Report of the National Food Survey Committee published by Her Majesty's Stationery Office as Household Food Consumption and Expenditure. 'Leafy salads' is synonymous with lettuce, the other salads quoted being tomatoes and cucumbers. This survey gives the annual consumption of salads in Britain as over 9 kg per head, of which 5.60 kg is tomatoes, 2.20 kg lettuce and 2.00 kg cucumbers.

The consumption of salads in Britain is seasonal and traditionally autumn and winter show a low demand for salads and spring and summer a high demand. This has to do with the English climate. There is a natural desire for salads in warm weather and a natural inclination towards cooked foods in cold weather. In Table 1 the seasonal, three monthly difference in salad consumption is shown against the opposite seasonal trend for potatoes, which of course are cooked. The marked seasonal differences in salad consumption are, however, becoming less. There is a strong demand for salads during autumn and winter which cannot be completely realised owing to the relative scarcity of supplies of quality salads over this period.

Seasonal consumption of salads shows a similar general pattern for all salads, the consumption of the salads taken together practically doubling in the spring period April to June over that of the winter quarter, January to March, to rise further in the summer period July to September and then to fall by half in the autumn and winter period,

Table 1

Salads consumption per head by season, ounces per week

	Jan.-Mar.	*Apr.-Jun.*	*Jul.-Sep.*	*Oct.-Dec.*	*Yearly average*
LETTUCE	0.75	1.92	2.57	0.81	1.51
per cent of year	12.4	31.7	42.5	13.4	
TOMATOES	2.38	4.18	5.46	3.09	3.78
per cent of year	15.8	27.7	36.1	20.4	
CUCUMBERS	0.59	1.42	1.47	0.57	1.01
per cent of year	14.6	35.0	36.3	14.1	
ALL SALADS	3.72	7.62	9.50	4.37	6.30
per cent of year	14.8	30.3	37.6	17.3	
POTATOES	42.05	34.14	36.07	47.00	36.90
per cent of year	26.5	21.4	22.6	29.5	

October to December. Individual salads show some variation in seasonal consumption over the pattern taken together.

The distribution of expenditure per head by season (see Table 2) reveals the quality and supply factor.

Table 2

Salads purchased by season, pence per week

	Jan.-Mar.	*Apr.-Jun.*	*Jul.-Sep.*	*Oct.-Dec.*	*Yearly average*	*£ per year*
LETTUCE	3.60	5.72	4.16	2.64	4.03	2.10
per cent	22.3	35.5	25.8	16.4		
TOMATOES	7.15	14.18	10.40	6.03	9.55	4.97
per cent	18.9	37.6	27.6	15.9		
CUCUMBERS	1.97	4.00	2.81	1.60	2.59	1.35
per cent	19.00	38.5	27.1	15.4		
ALL SALADS	12.82	24.90	18.37	10.27	16.17	8.42
per cent	19.3	37.5	27.7	15.5		

In April and June, expenditure, like consumption, is double the amount spent in January and February. However, expenditure in July and September falls as consumption rises, to fall further in October to December than in the first three months of the year. July to September is the period of the highest domestic supply when the price of all salads on the market falls.

These quarterly and yearly consumption and expenditure figures are affected by the number of households purchasing the produce.

From Table 3, it can be seen that the seasonal percentage of households purchasing salads follows a different pattern from consumption and expenditure. The number of households purchasing salads is highest in all cases in the April to June period, i.e.,

Table 3

Per cent households purchasing salads

	Jan.-Mar.	*Apr.-Jun.*	*Jul.-Sep.*	*Oct.-Dec.*	*Yearly average*
LETTUCE	25	46	38	23	33
TOMATOES	42	61	53	35	48
CUCUMBERS	18	35	29	15	24

higher than in the July to September period, when consumption and expenditure is greatest. This suggests that more households buy at the peak of spring than at any other time. This may be because more produce comes from home gardens in July to September.

Prices over the season are very much a consequence of supply, quality and demand (Table 4).

Table 4

Prices over the season, pence per lb., 1984

	Jan.-Mar.	*Apr.-Jun.*	*Jul.-Sep.*	*Oct.-Dec.*	*Yearly average*
LETTUCE	77.7	53.6	33.6	54.8	49.9
TOMATOES	49.8	54.6	42.7	44.3	48.3
CUCUMBERS	53.2	45.2	36.8	46.0	48.3

Lettuce prices, for instance, are highest from January to March, more than double those in the period July to September and considerably higher than those in the three months before, namely October to December. Tomato prices are more equable over the year but in the first six months they are much higher, reaching a maximum in April to June (spring), than in the second half of the year. Cucumber prices are highest in the first three months and lowest from July to September. This situation is due to the fact that there are no relatively cheap imports of lettuce in the off-season of domestic supply. The dominant supply of tomatoes in the off-season of North European supply is that from the Canaries, of relatively inferior quality and lower price, but for which there are no ready substitutes.

Finally, the response of the salads market to increasing incomes is discussed briefly. All salads have a good response to increases in the income per head. The income elasticity of expenditure or the extra expenditure for a one per cent rise in income, is 0.58 for lettuce, i.e., a 58 per cent increase in expenditure for a given rise in income. That for tomatoes is 46 per cent and for cucumbers it is higher at 67 per cent. The response for the quantity purchased for a given rise in income in 42 per cent for lettuce, 39 per cent for tomatoes and 39 per cent for cucumbers. The level of real incomes in Britain is of the greatest significance to the market for salads. Even now a large proportion of the population do not buy salads and the existing demand for salads, increasing as it is, will be still greater as incomes rise. Since 1971, real income in Britain has risen by 22 per cent and since 1974 real disposable income has risen by 15 per cent. This trend is most likely to continue and it will be much to the advantage of the salad market in Britain as its expansion will be assured. Another development associated with increases in income and disposable income of importance for the salad market is the number of households with a freezer. Fifty per cent had a home freezer in 1980 and 71 per cent in 1985. Rather more households have normal refrigeration which increases the market for salads.

LETTUCE: SUPPLY AND DEMAND

Figure 1 shows the supply of lettuce from 1971 to 1985, during which time the supply to Britain increased by 45 per cent.

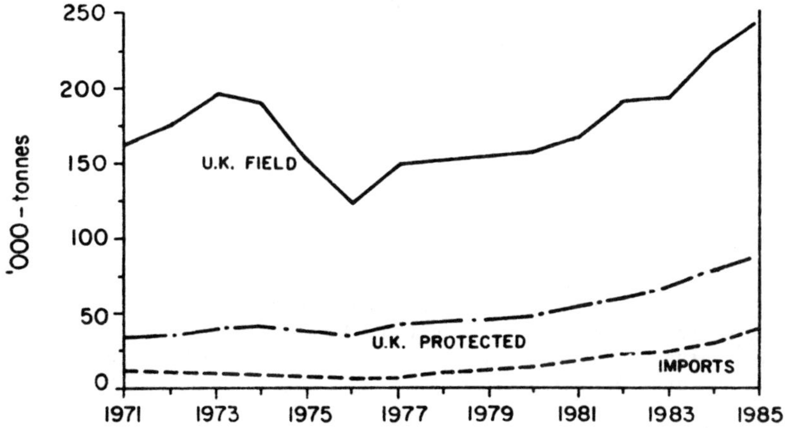

Figure 1 Lettuce supplies to the United Kingdom, 1971 - 1985

Home-grown field lettuce, which is available in quantity from June to September, dominates the supply over the years. It is from two to three times greater in supply than protected lettuce. At the market level, protected lettuce may almost be considered as a separate commodity, though field lettuce is a strong substitute for protected lettuce in its own season (summer). Field lettuce supplies, i.e., home-grown, after a peak in 1973, dropped considerably in the drought years of 1975 and 1976, from which they recovered slowly from year to year to pass the 1974 peak in 1984 and 1985.

Protected lettuce was 21 per cent of the lettuce supply in 1971 and it has risen to become 34 per cent of the supply in 1985. This represents an important shift in demand for a preferred commodity, made possible by increases in disposable incomes in Britain. Protected lettuce is made up of imports and home-grown. In 1971 about a third of protected lettuce supplies were imported; only a sixth was imported in the years 1975 to 1978, since when imports have risen steeply from a quarter of the supply in 1979 - 80 to become a third of the protected lettuce supply in 1980, and to hold this position in 1985, despite the steep rise in the supply of field lettuce since 1980.

It is clear that protected lettuce has reached a new threshold in consumption which is accompanied by a strong demand for field lettuce too. It would seem that the British public have become lettuce conscious. This is borne out by the figures of apparent consumption per head (total supply divided by the population). In 1971 field and protected lettuce consumption together was 3 kg per head, increasing to 4.24 kg per head. The consumption of protected lettuce more than doubled.

The changes in consumption per head are given in Table 5.

Table 5

Changes in consumption per head, kg

	Protected lettuce	*All lettuce*
1971	0.62	2.96
1973	0.63	3.46
1975	0.66	2.71
1977	0.70	2.63
1979	0.78	2.70
1981	0.91	2.93
1983	1.09	3.33
1985	1.45	4.24

Following an initial high level, there was a decline in the mid-seventies associated with drought (in 1975 and 1976), then a gradual return to higher consumption, with a steeper increase in the 1980s at a sustained and increasing level.

Imports and home supplies in 1985

Britain imports thirty-four thousand tonnes of lettuce. The category includes cabbage lettuce, chicory and endive. Monthly imports to Britain are shown in Figure 2; they comprise twelve thousand tonnes from other countries in the EEC and twenty-two thousand tonnes from third countries.

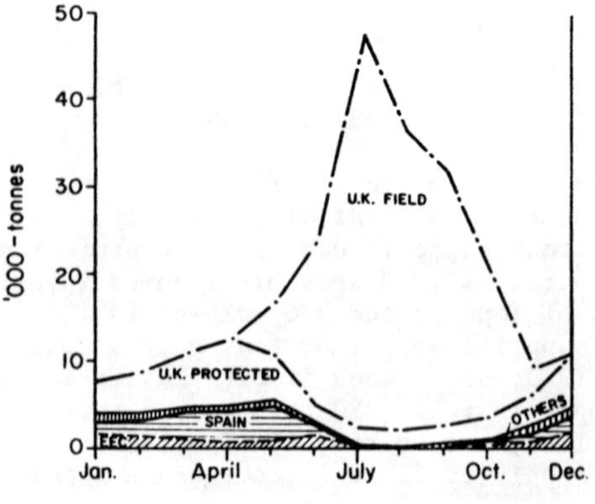

Figure 2 Monthly supplies of lettuce in the United Kingdom in 1985

In 1985 Spain was outside the EEC. Of the twelve thousand tonnes imported from EEC countries, no less than nine thousand tonnes came from Holland and of third country imports, twelve thousand tonnes came from Spain. Following was Israel with thirteen hundred tonnes and next, indicating the value of lettuce in world trade, the United States, with twelve hundred tonnes. In Europe minor imports came from France, a thousand tonnes, and a similar amount from Italy. Ireland which to a very slight degree is the kitchen garden of England, sent two hundred and fifty tonnes and Portugal, in 1985 a third country, sent five hundred tonnes.

Imports are greatest in May, at 15 per cent, and reflecting the home supply, they are low from July to October. In July only one per cent of the imports come in, in August half that amount, in September 1.3 per cent and in October three per cent. Imports build up in November to eight per cent, and in December to almost double this figure. Imports in the New Year start at ten per cent of total imports for January and increase each month to a peak of 15 per cent in July. Thus the main import season is from November to June when home supply is either scarce or absent. EEC, north temperate imports, come all the year round and are high from November to June, which is also the pattern of third country imports. Imports from Spain start in November at over a thousand tonnes, reaching over three thousand tonnes per month in March and four thousand in May till the first week in June, when they decline until the last week in October.

Countries from all over the world (Africa, the Middle East and South East Asia) send small consignments of lettuce to Britain. Even Iran sends eleven tonnes evenly over the months from January to April. The United States enters the market in October, in common with most imports and sends over the winter to the beginning of May, never sending more than two hundred and fifty tonnes in one month. Israeli imports show the same pattern, increasing from one hundred and forty tonnes in December to four hundred and fifty tonnes in February, then decreasing in March and ending in April.

Monthly market prices

Table 6 shows the monthly market prices for three types of lettuce (home-grown and imported) - cabbage, crisp including Iceberg and Iceberg. Prices of home-grown cabbage lettuce are lowest for field lettuce, which is quoted from June to November. The price of home-grown protected lettuce is not appreciably higher over this period, except for the beginning and end of the June to November period. From June to November the price of cabbage lettuce from Holland (and Belgium) is substantially higher than that of home-grown, a premium of some 20 per cent or more. Dutch cabbage lettuce has a good premium on home-grown throughout the year. The price of home-grown crisp, including Iceberg, are only listed for April and May. All imports from January to May have a price two or three times that of the home category. The situation is different for Iceberg lettuce. Overall the home-grown product demands similar prices to imports, except at the beginning and end of the season. It is clear that the home-grower can get quality equal to imports in his natural growing season.

Table 6

Lettuce - Monthly prices 1985, £ per dozen

	Jan.	*Feb.*	*Mar.*	*Apr.*	*May*	*Jun.*	*Jul.*	*Aug.*	*Sep.*	*Oct.*	*Nov.*	*Dec.*
Cabbage lettuce												
Home grown												
average all classes, field grown						1.16	1.10	0.92	1.14	1.01	1.08	
average all classes, protected	1.81	2.38	2.13	1.34	1.29	1.53	1.18	1.00	1.24	1.11	1.28	1.82
Imported												
Belgium, all qualities										1.88	1.72	
Netherlands, all qualities	2.48	3.25	2.82	2.15	1.85	1.79	1.55	1.39	1.65	1.52	1.79	2.49
Crisp, including Iceberg												
Home grown												
average all classes				2.34	2.19							
Imported												
Israel, all qualities	6.92	5.88	6.27	6.18								
USA, all qualities		6.58	5.48									
Spain, all qualities	5.76	4.93	4.66	4.74	4.20							
Iceberg												
Home grown												
average all classes						4.57	3.31	3.04	3.17	3.28	4.61	
Imported												
Spain, all qualities						5.56				3.03	7.08	6.10
Netherlands, all qualities							4.42			3.76		6.56

TOMATOES: SUPPLY AND DEMAND

Figure 3 shows the supply of tomatoes from 1971 to 1985, during which time the supply to Britain increased by 17 per cent, mostly during the last four years.

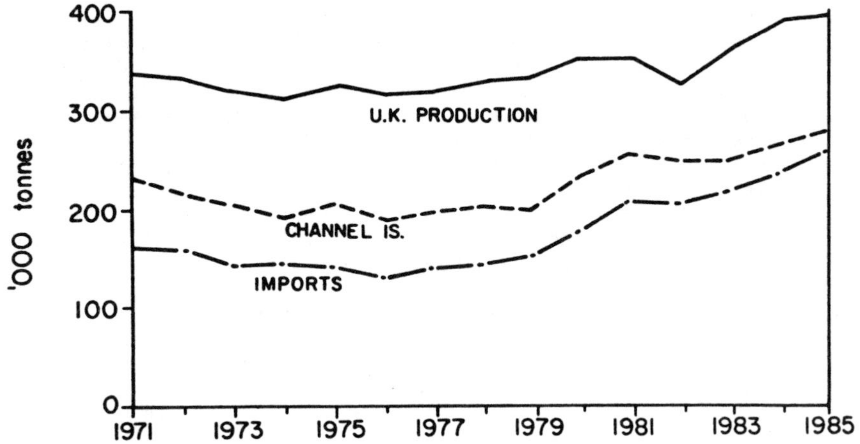

Figure 3 Tomato supplies to the United Kingdom, 1971 - 1985

However, the proportion imported has markedly increased from 48 per cent of the supply in 1971 to 64 per cent in 1985. This is important to countries considering exporting to Britain.

Tomatoes consumed in Britain are predominantly of two kinds, the English and North European heated glasshouse grown tomato, of which that of the Netherlands is of overwhelming importance, and the Spanish and Canary type. Spanish tomatoes are grown with no heat in a Mediterranean climate, and though in certain areas they are grown under plastic structures, vast areas of tomatoes are grown in the open. Canary tomatoes are almost exclusively grown in the open in a sub-tropical climate. Most of the Spanish supplies arrive in the last quarter of the year, October to December inclusive, after which tomatoes from the Canaries take over in force to the end of May.

The North European glasshouse grown supply is available in quantity from as early as April to the end of October, throughout the English summer when demand for tomatoes is high. It is complementarity of climate and the resulting harvest time in different climatic zones which gives Britain a supply of tomatoes all the year round. This also applies of course to lettuce, cucumbers and celery and the much less significant salad produce including sweet peppers. North European supplies, however, represent the preferred, choice tomato, of high quality. These supplies are also grown in close proximity to the British market, which implies not only the very important feature, freshness, but also relatively low transport costs. Packaging and presentation are necessary expenditures for meeting the competition.

Mediterranean tomatoes do not attain such a high standard of quality unless they can be grown under the same exacting environmental conditions as the North European glasshouse grown tomato. However, there is no substitute for them for most of the autumn and spring season and they bear a competitive edge in being cheaper than the North European tomato.

As in any country, the market is sectioned according to the level of income, and a significant proportion of the purchasing power in Britain cannot afford the best at the highest prices. Market demand is for tomatoes throughout the year and competition in quality, from whatever source, is a strong factor. A sandstorm in the Canaries, for instance, can cause a dislocation in supply for a matter of weeks. The more distant suppliers, of course, carry a necessary transport cost penalty but this is more than offset by savings in energy that would be incurred for growing the produce. In an ideal world, produce responsive to the sun's energy would be supplied to areas of the world which lack such abundance. It is, however, the very point of quality which gives the North European tomato the pre-eminence in the market - people will pay more for it. The same goes for lettuce and for cucumbers. At the margins of the field lettuce season in Britain the premium on protected lettuce is greater, but decreases with the increase in quality and availability of field lettuce in summer. This ideally coincides with the season when demand for lettuce is greatest.

As shown in Figure 3, the demand for tomatoes, indicated by the supplies line, declined from 1971 until 1980, when it recovered to increase by one sixth (40,000 tonnes) between 1980 and 1985. The slack years, like those for lettuce but over a longer period, are most likely explained by pressure on incomes during those years, with some resurgence as people become more aware of the importance of salads to health. The increasing share of imports over the period 1971 to 1985 is readily apparent. This increasing share has of course been accompanied by static or near static home production over many years, which rose for some years before 1978, since when it has dropped by almost a quarter, Channel Islands production, nearly two fifths of home tomato supplies in 1971, fell by half to become a fifth of home supplies in 1985. Both the fate of the glasshouse tomato industry in the Channel Islands and the near freezing to a static position of the glasshouse industry in Britain is a measure of the successful competition of Holland, which of course markets tomatoes during the British season.

Apparent consumption in 1971 was 6 kg per head and it did not reach beyond this figure until 1980, when it fell in one year to only 5 1/2 kg. Since 1980, consumption has shown a rapid and sustained increase, reaching 7 kg per head in 1985.

Imports of tomatoes in 1985

Imports of tomatoes from the main sources are shown in Figure 4, which illustrates that the countries shown supply virtually all the imports of tomatoes to Britain. Forty-four per cent or one hundred and eleven thousand tonnes comes from countries in the EEC, with over one hundred thousand tonnes from Holland.

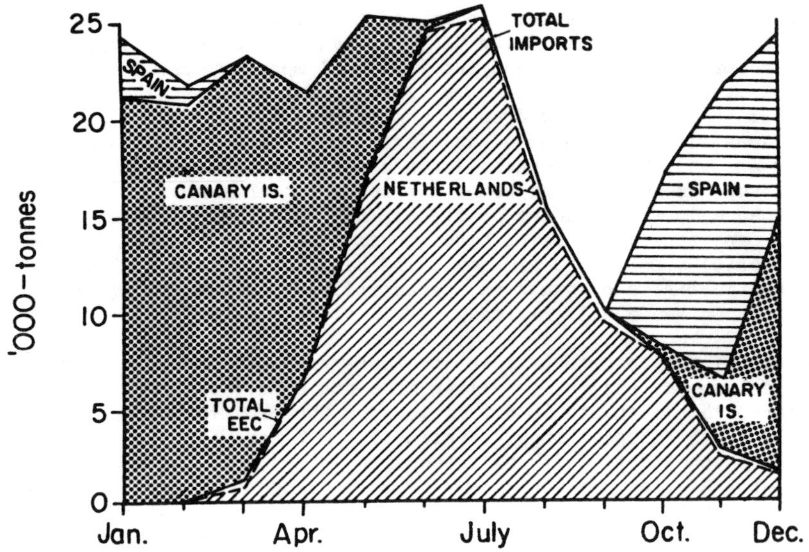

Figure 4 Monthly imports of tomatoes into the United Kingdom in 1985

The other significant supplier, but with only two thousand tonnes, is the Republic of Ireland, which traditionally has looked to Britain as a country to which to export, in order to help make its small horticultural sector viable. France, Belgium and even the Federal Republic of Germany send very small consignments. Of course, all North European imports come in the British summer season, but it should be noted that Italian exports of tomatoes are so small as to be practically nonexistent, and those from Greece are even smaller. There seems to be an unwritten cleavage such that Italy exports to the Federal Republic of Germany while Spain exports to Britain. This no doubt has arisen out of tradition and geographical location. Britain has had strong trading relationships with Spain for centuries and Spain has traditionally supplied as much as one sixth of Britain's imports of fruit and vegetables. Spain entered the EEC and its new relationship with Britain on this basis. In 1985 it should be noted that Spain was still a third country, qualifying for membership of the EEC on 1 January 1986. Spanish entry into the EEC will of course have great consequences, especially over the course of the next five to ten years, as Spain reaches the end of her EEC transitional period. This will be of importance to Mediterranean third countries in the export of and competition in the market for fruit and vegetables, both for trade with the EEC as a whole and for trade with Britain. The consequences for Mediterranean third countries does of course depend on EEC policy towards them, which is now under negotiation. It is unlikely for strategic reasons which affect the EEC's own interests that the EEC will adopt a policy detrimental to the Mediterranean third countries concerned.

The total supply from the third countries in 1985 was one hundred and forty-two thousand tonnes of tomatoes, of these no less than one hundred and four thousand tonnes came from the Canary Islands and the next most important supplier, Spain, sent thirty-seven thousand tonnes. Though Spain is now part of the EEC, the Canaries are not. They prefer to be an important tourist and international monetary centre. Although fourteen other third countries are listed separately as exporting about one thousand five hundred tonnes between them, Morocco is conspicuous with seven hundred and forty tonnes and Israel with three hundred and forty-three tonnes. Token exports of tomatoes to Britain in fact come from most regions of the world but they have no significance in true trading practice. They do not even represent serious opportunist marketing since consignments are unrepeated over the months of supply, and are small and haphazard. These include Iceland, the USA, and countries in Central and South America, in the Caribbean, Oceana and in Africa.

Imports of tomatoes to Britain are lowest in September, towards the end of the summer season, totalling nine and a half thousand tonnes. In this month all tomato imports come from the EEC. In October, as the home and North European supply declines, the Canary Islands and the Spanish season starts, with nine thousand tonnes from Spain and a two hundred tonnes from the Canaries. November brings more from Spain at fifteen thousand tonnes and even more from the Canaries at three and a half thousand tonnes. Imports from the Canaries increase rapidly in December to fourteen thousand tonnes, while the deliveries from Spain fall to nine thousand tonnes. North European imports, which as mentioned reach nine and a half thousand tonnes in September, fall off each month to December, when they run at a little over the thousand tonne mark. These imports remain much below a thousand tonnes per month until March. From January to March Canary Islands exports reach their peak of twenty thousand tonnes a month, displacing those from Spain, whose exports, at three thousand tonnes in January, decline steeply in February and March, from which months they are only nominal. Morocco is in the market from November to April and Israel from January to March. The amounts, however, are small, Morocco exporting seven hundred and forty tonnes to Britain and Israel three hundred and forty tonnes.

From April to July North European imports climb progressively each month to reach between twenty-five and twenty-six thousand tonnes in July. Canary Islands imports fall to fourteen thousand tonnes in April and to eight thousand in May, with a minor consignment in June marking the end of the Canary Islands season. Small amounts of Spanish tomatoes, less than one hundred tonnes per month, arrive until July.

The European supply of tomatoes (mainly Dutch) to Britain emerges from its winter low in March, merely a thousand tonnes, to become six and a half thousand tonnes in April and sixteen thousand tonnes in May. In June the supply increases quickly to twenty-four thousand tonnes. The European supply then progressively falls from twenty-five thousand tonnes in July to nine thousand in September and one thousand three hundred tonnes in December. The Dutch exporters are geared to producing as early as possible and they now export a significant quantity in March, to gain the high prices in the early season. They continue to export as much as possible by the end of June, while high prices last. Prices progressively fall from July onwards.

The tomato supply is of course supplemented by the important home crop which meets the bigger summer demand and offers tomatoes to a discriminating and price conscious public alongside and in competition with the Dutch. During other seasons the competition is straight between Spain, the Canaries and other Mediterranean countries which, at the margins of their seasons of supply, also have to compete strongly against the North European supplies. Competition at all seasons is essentially for the market for quality produce. The complementarity of home supplies of tomatoes and imports is shown in Figure 5.

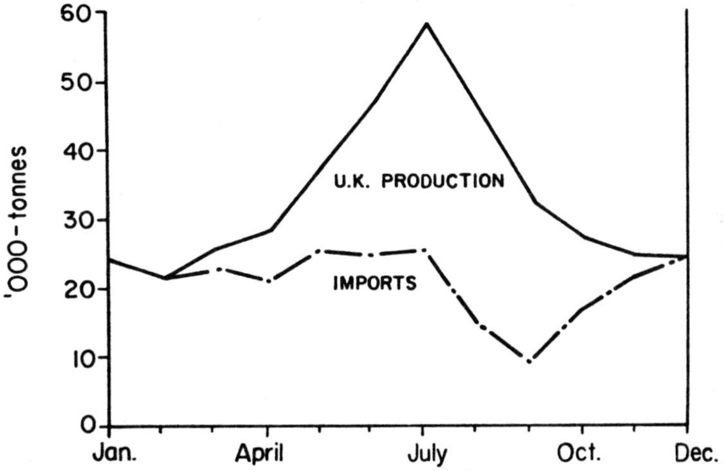

Figure 5 Monthly supplies of tomatoes in the United Kingdom in 1985

A few large nurseries in Britain with a large area have shifted to the Dutch pattern of early production, to put home supply earlier in the season than before, but the bulk of home production comes later, including the cold grown crop which finishes late in the autumn.

In 1985 the home crop amounted to one hundred and forty-three thousand tonnes. This shows how important tomatoes are in the family nursery which grows several glasshouse crops. These nurserymen do not go for the more profitable early crop because they cannot provide the high investment required and indeed it may well not be profitable for them.

The tomato market may be summarised as follows:

Supply	*Tonnes*	*per cent of imports*	
		Year	*Dec. - Mar.*
Home production	143,400		
EEC imports	111,042	43.8	3.0
Holland	107,936		
Ireland	2,095		
Third country imports	142,452	56.2	97.0
Spain	37,086		
Canary Islands	103,858		

Monthly market prices

The price of home-grown tomatoes from February to May is well above that of Canary Islands tomatoes since in this period home-grown tomatoes are relatively scarce and very scarce in the first few months (see Table 7).

Table 7

Tomatoes - Monthly market prices 1985, £ per 6 kg

	Jan.	*Feb.*	*Mar.*	*Apr.*	*May*	*Jun.*	*Jul.*	*Aug.*	*Sep.*	*Oct.*	*Nov.*	*Dec.*
Home-grown average all classes		9.70	7.10	6.37	4.97	4.64	2.61	1.92	2.01	1.99	1.93	
Imported Canary Islands, all qualities		3.34	3.80	4.10	3.91	3.43				2.63	2.68	4.28
Morocco												
Israel, all qualities			2.93									
Netherlands, all qualities			6.91	6.49	4.81	4.43	2.62	2.02	2.31	2.68	2.80	
Spain, all qualities		2.27								2.71	2.82	4.52

The price of home-grown tomatoes falls greatly in July to almost half the June price, but poor quality in October and November gives the new Canary Islands tomatoes a premium over home-grown. This also applies to supplies from Spain from October to the end of the year. Dutch tomatoes, available over much of the year, fetch high prices, but not quite so high as the English tomatoes from March to June, but they do better in price from June onwards.

The Dutch in the tomato trade

Because of the importance of Holland in the tomato market, it is of value to observe Dutch trade in this market. Figures for three recent years are given in Table 8. Dutch exports are on a rising trend. The chief market is West Germany, followed some way behind by Britain and France, with Scandinavia as an important smaller market in proportion to their population.

From the Dutch point of view the total sales to the end of June, the most lucrative part of the market season, is of great importance. From the point of view of third country Mediterranean exporters it is most important to measure the market in a given country at any one time to

Table 8

Dutch tomato export trends, 1983 - 1985

		Mar.	*Apr.*	*May*	*Jun.*	*to end Jun.*	*Jul.*	*Aug.*	*Sep.*	*Oct.*	*Total*	*Indices (1832=100)*
West Germany	1983	912	17,996	31,475	43,001	93,394	50,084	30,279	17,524	15,599	206,880	100
	1984	2,088	17,440	27,950	41,157	88,635	42,715	36,649	34,525	13,367	215,885	104
	1985	2,860	19,201	30,580	36,296	98,937	52,636	35,356	31,007	12,290	225,226	106
UK	1983	40	4,156	15,342	20,830	40,368	23,603	9,292	5,546	567	84,426	100
	1984	331	6,991	16,385	22,711	46,418	22,140	10,607	6,606	4,138	89,909	106
	1985	445	6,405	12,644	22,273	41,767	27,636	14,636	11,554	6,159	101,752	121
France	1983	15	3,946	7,764	14,222	25,947	9,259	3,930	5,948	5,477	50,561	100
	1984	75	6,297	8,017	13,221	27,610	14,448	8,202	4,421	2,473	57,154	113
	1985	544	6,381	10,992	12,810	30,727	15,304	6,385	4,197	2,653	59,266	117
Scandinavia	1983	156	3,759	3,971	4,465	12,351	3,928	2,388	1,714	1,847	22,228	100
	1984	311	4,393	3,701	3,423	11,828	2,675	2,044	2,830	1,765	21,142	95
	1985	1,061	5,465	3,974	3,869	14,369	3,764	3,091	2,615	1,676	24,915	112
Other	1983	475	2,964	3,646	3,345	10,430	2,884	209	1,971	2,640	18,134	100
	1984	638	3,485	3,748	4,659	12,530	3,551	675	1,225	2,188	20,169	111
	1985	390	3,932	5,284	4,981	14,587	3,952	1,069	3,284	3,470	26,362	145
Total	1983	1,598	32,821	62,198	65,873	182,490	89,758	46,098	32,703	31,180	382,229	100
	1984	3,443	38,606	59,801	85,165	187,021	85,529	58,177	49,607	23,931	404,259	106
	1985	5,300	41,384	63,474	80,229	190,387	102,692	60,537	52,657	26,248	432,521	113

Distribution of Dutch exports, per cent

	1983	1984	1985
West Germany	54.12	53.39	50.92
UK	22.09	22.23	23.53
France	13.23	14.13	13.70
Scandinavia	5.81	5.23	5.76
Other	4.75	4.99	6.09

Exchange rates, Sterling

	West Germany	Holland	France
1983 June	3.95	4.70	11.87
1984 June	3.77	4.25	11.59
1985 June	3.92	4.42	11.94
Sterling movement since 1983	-1%	-6%	+1%

Source: Dutch Bureau of Auctions, 1986.

see what part of the supply they can fill. The Dutch are formidable competitors and other exporters must have a product either of equal quality or a good substitute for the best quality at a lower price.

CUCUMBERS: SUPPLY AND DEMAND

Figure 6 shows the supply of cucumbers from 1971 to 1985, during which time the supply to Britain increased by 60 per cent. Of great significance is that over this time imported cucumbers doubled in quantity from a share of 30 per cent to one of 60 per cent of the total supply. Consumption is continuing to rise.

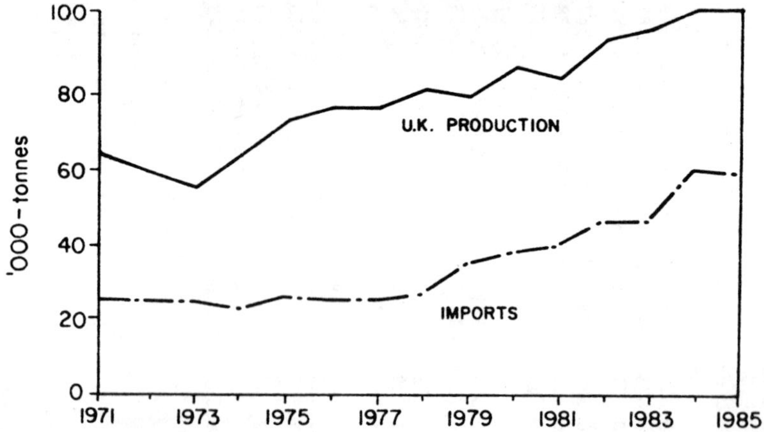

Figure 6 Cucumber supplies to the United Kingdom, 1971 - 1985

Referring to Figure 6, it is apparent that imports of cucumbers have always been important in the supply to Britain. In 1971 when total supplies were sixty-four thousand tonnes, twenty-five thousand tonnes were imported. Imports remained more or less stable to 1977, when they were still twenty-five thousand tonnes, but home supplies had increased from thirty-nine thousand tonnes in 1971 to fifty-one thousand tonnes. The year 1977 marked a take-off point for the increase in imports of cucumbers to Britain. By 1983 supplies of cucumbers reached over one hundred thousand tonnes for the first time, of which forty thousand tonnes were imported. Home production of cucumbers is now at a level of fifty to fifty-five thousand tonnes, whilst imports are now almost equal, totalling forty-eight thousand tonnes in 1984 and forty-seven thousand tonnes in 1985.

Apparent consumption of cucumbers in 1971 was 1.14 kg per head, reaching 1.82 kg in 1985. Consumption reached a low point in 1973, 0.98 kg per head, since when it has climbed steadily upwards to double the 1973 figure. It is clear that since 1981 when disposable income in Britain have been rising at a significant rate, cucumbers have shared the recent boost in consumption with other salads, notably tomatoes and lettuce, and celery in the last three years.

Cucumber imports in 1985

North European glasshouse grown supplies, almost entirely from the Netherlands, dominate imported supplies of cucumber to Britain more so than tomatoes (see Figure 7), but the Canary Islands and Spain play an important part.

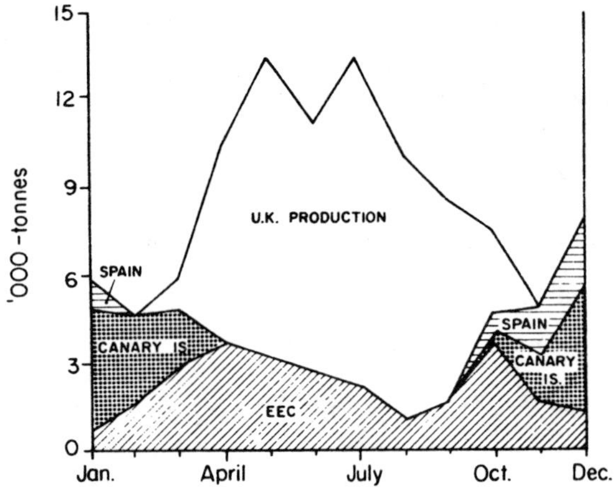

Figure 7 Monthly supplies of cucumbers in the United Kingdom in 1985

In 1985 forty-seven thousand tonnes were imported, twenty-six from Holland and twenty thousand tonnes from third countries, almost all from the Canary Islands and Spain. The Netherlands' cucumber exports have a different supply pattern compared with tomatoes, and are in the market more consistently over the year, but more heavily from February to July. The Netherlands send two to four thousand tonnes in each of these months, and subsequently substantial quantities from September (inclusive) and later. The Canary Islands and Spain, sending fifteen thousand and five thousand tonnes, respectively, over the year, start their season in October, the month when Holland sends over three thousand tonnes. The Canary Islands send four hundred and ninety-five tonnes in October, and thousand six hundred in November and four thousand in December. January is the peak month, with four thousand four hundred tonnes. This level falls to two thousand five hundred tonnes in July and one thousand eight hundred tonnes in April, followed by a trickle over the next three months. Spanish exports comprise five hundred and seventy tonnes in October, one thousand six hundred tonnes in November, two thousand tonnes in December, and seven hundred and forty tonnes in January after which the supply falls off sharply. As with tomatoes, Holland dominates the European cucumber market. This success is as much to do with her auction system as her technical prowess, but cucumbers do give her a year round trade which helps to put export of cucumbers, which are of less value than tomatoes in total, on a viable footing. As with tomatoes, any potential exporter to Britain must give full attention to what the Dutch can do in this market. There must be room in any one month for the export and it must be competitive with the Dutch, or at least provide a quality substitute at a lower price.

Cucumber exports to Britain from other countries do not exceed one hundred tonnes. Here Cyprus is the strongest contender, sending over ten tonnes per month from December to April, coinciding with exports from Spain and the Canary Islands. All other countries dispatch some twenty tonnes or less but their attention to the British market is interesting. Egypt sends eleven tonnes in January when Holland transports her lowest consignment of six hundred and seventy-five tonnes. Israel also sends her total eleven tonnes in January and Jordan sends five tonnes in January and fifteen tonnes in March. Rumania, an example of a country in need of hard currency, sends twenty-five tonnes, against four from Belgium and Luxembourg and over three thousand from Holland.

Monthly market prices

Monthly market prices for cucumbers are shown in Table 9.

Table 9

Monthly market prices 1985, £ per 5 kg

	Jan.	*Feb.*	*Mar.*	*Apr.*	*May*	*Jun.*	*Jul.*	*Aug.*	*Sep.*	*Oct.*	*Nov.*	*Dec.*
Home-grown average all classes		4.79	3.97	3.36	2.62	2.39	1.56	1.29	1.33	1.68	1.82	
Imported												
Canary Islands, all qualities	3.36	4.79	3.76	2.95						3.40	2.84	4.44
Greece, all qualities	3.07	4.60										
Netherlands, all qualities		5.31	4.85	4.05	3.40	3.05	2.19	1.93	2.29	3.27	3.61	
Spain, all qualities	3.05									2.77	2.66	4.11

Here a comparison between English cucumbers and those from Holland reveals a pattern reverse of that for tomatoes. Dutch cucumbers have the edge in price on English cucumbers over the whole year with a market premium at the end of the year. Canary Islands prices are similar to the English grown cucumbers at the beginning of the year, decreasing slightly in April. At the end of the year they are much higher in price than the English cucumber. Spain and France export for a short period in the high priced winter season but their prices to not match those of the Dutch.

CELERY: SUPPLY AND DEMAND

In 1971 the celery supply to Britain was sixty-six thousand tonnes, of which only nine thousand tonnes were imported (see Figure 8).

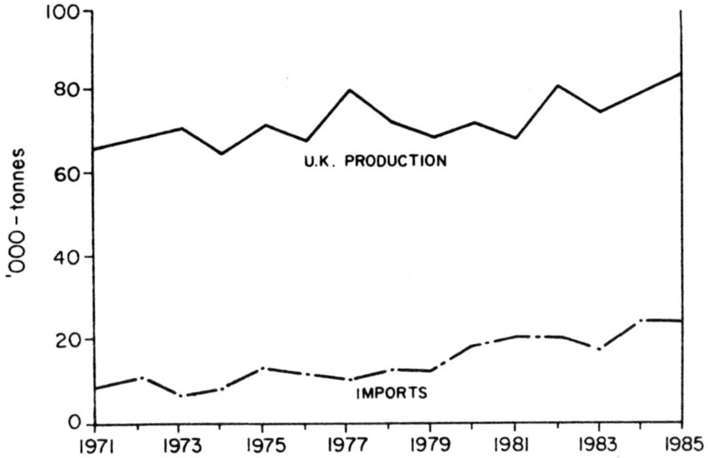

Figure 8 Celery supplies to the United Kingdom, 1971 - 1985

Home supplies were strong, reaching over sixty thousand tonnes in 1973 and nearly seventy thousand tonnes in 1977, to fall off slightly in the 1980s, but the home supply remains important and dominant in the market for celery. In recent years home-grown protected celery, which is on the increase, has reached some nine thousand tonnes.

In 1985 the total celery supply increased over that in 1971 by twenty thousand tonnes - an increase in the market of almost a third. The expansion has been in protected celery, mainly imported. Since 1971 imports of celery have nearly trebled, reaching twenty-five thousand tonnes in 1985, doubling the twelve thousand tonnes of 1979. The expansion in demand, affecting imported celery in particular, which has been evident in the last five years, is likely to continue. In terms of apparent consumption, demand from 1971 to 1976 was some 1.2 kg per head. In 1975 and 1982 it reached over 1.4 kg per head but from 1983 to 1985 it rose above the general trend to become 1.32 kg in 1983, 1.41 in 1984 and 1.53 in 1985. The outlook for the celery market in Britain is promising. Home-grown supplies of field and protected celery will be important but the demand for celery in the off-season of home supply is increasing. This gives exporters a great opportunity.

Celery imports in 1985

Imports of celery to Britain in 1985 were twenty-four thousand six hundred tonnes, much less than the other salads considered here, but nonetheless an important import (see Figure 9).

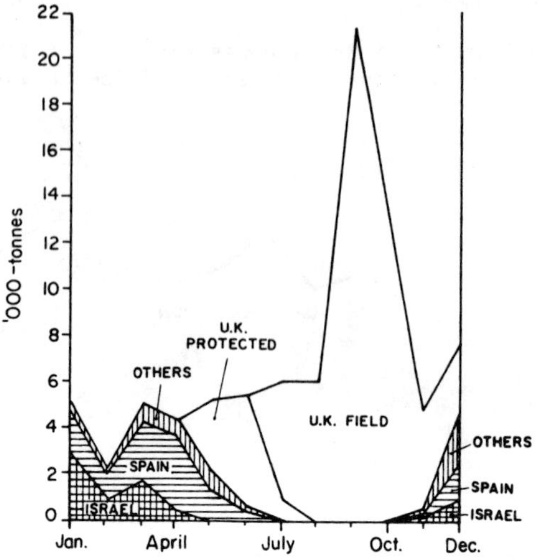

Figure 9 Monthly supplies of celery in the United Kingdom in 1985

In contrast to the other commodities, imports of celery come substantially from third countries, totalling 80 per cent, two thirds of which come from Spain. EEC imports mainly come from the Netherlands at two and a half thousand tonnes, from France with one thousand six hundred tonnes and from Italy at six hundred tonnes. More than a third of non-EEC imports come from Israel. From EEC countries, imports begin in November at one hundred and ten tonnes, and are highest in the following month at two thousand tonnes, over half coming from France. Some 45 per cent of EEC imports arrive in December. EEC imports are considerably lower in January and February, rising to 15 per cent of all imports in March and they remain at this level or above until May, falling off in June as the home crop is put on the market. The home crop has the market from June until November or December.

Third country imports in November are three hundred and fifty tonnes, to become one thousand six hundred and fifty tonnes in December, the month when marketable home supplies become scarce with the incidence of winter frosts. Third country imports, almost all from Spain and Israel, then continue to dominate the market until June. Israeli supplies peak from December to January, when they are almost three thousand tonnes, and continue to be high till April, when they fall off for the summer season. The other third country exports of significance are from the United States with four hundred and thirty tonnes, which come in from November to May when the home crop is not available.

Monthly market prices

Celery prices are shown in Table 10. No prices are quoted for imports from July to the end of the year, which indicates that the English celery then has the market to itself. The period of imports from January to May - June is of interest. Prices are higher, but Israel has the edge in price over supplies from Italy and Spain. At the same time the English protected crop demands higher prices than imports in April and May on account of quality and freshness.

Table 10

Celery - Monthly market prices, £ per 12 pieces

	Jan.	*Feb.*	*Mar.*	*Apr.*	*May*	*Jun.*	*Jul.*	*Aug.*	*Sep.*	*Oct.*	*Nov.*	*Dec.*
Home-grown												
self-blanching, average all classes					3.55	3.77	2.66	2.12	1.81	1.75	1.88	2.50
Main crop, unwashed average all classes			1.96								2.09	1.99
Main crop, washed average all classes								1.95	1.83	1.95	2.21	2.41
Imported												
Israel, all qualities	3.07	3.14	3.68	3.49	3.44							4.08
Italy, all qualities	2.82			3.36	3.24							4.09
Spain, all qualities	2.65	2.58	3.29	2.79	2.94	3.10						3.74

EEC POLICY AND SALADS

Spain and Portugal were disfavoured by EEC policy as compared with most Mediterranean countries, before their accession. For instance, the Common Custom Tariff on tomatoes was 11 per cent for Spain, but 4.4 per cent for Morocco, Algeria, Tunisia, Egypt, Cyprus and Malta, while the CCT on peppers was 6.3 per cent for Spain and 5.4 for Morocco, Algeria, Tunisia, Israel, Cyprus and Malta, with 4.5 per cent for Turkey. The result of Spanish and Portuguese accession will be to put them in a completely favourable position as members of the EEC and other third country Mediterranean countries, though they now stand to get better EEC accessions, will be relatively disadvantaged.

The policy of the EEC in the fruit and vegetable sector stands mainly on two legs: on the Reference or Minimum Import Prices and on the Common Customs Tariff. These apply to the EEC's external policy. EEC reference prices for salads are shown in Table 11.

Table 11

EEC reference prices or minimum import prices

LETTUCE	Reference price period 1 November to 31 May	
		ECU per 100 kg
	1 November - 31 December	67.39
	1 January - 28 February	72.40
	1 March - 31 May	80.04
ENDIVE	Reference price period 1 November to 31 March	
	1 November - 31 January	57.16
	1 February - 31 March	61.40
TOMATOES	Reference price period 1 April to 20 December	
	1 April - 30 April	197.27
	1 May - 31 May	136.75
	1 June - 10 July	99.96
	11 July - 31 August	41.06
	1 September - 30 September	44.43
	1 October - 20 December	45.53
CUCUMBER	Reference price period 11 February to 10 November	
	11 February - 20 February	142.67
	21 February - 28 February	122.42
	1 March - 31 March	112.14
	1 April - 30 April	96.76
	1 May - 31 May	76.12
	1 June - 30 June	63.76
	1 July - 31 July	47.29
	1 August - 31 August	47.66
	1 September - 30 September	56.42
	1 October - 10 November	80.79

The ECU is an EEC value - in July 1986 1 ECU = 64 p. The minimum import prices are stepped according to season to give the highest protection in the season of EEC supply, rather like the former UK tariff system. Thus, tomatoes have the highest reference price in April, to fall by almost a third in May and by another third in July, when it levels out at a lower amount for the rest of the season. Lettuce, not so protected, has reference prices for three periods, differing in amount. Cucumber, like tomatoes, have the highest reference price at the beginning of the season, with a lower one for the months towards the summer. The reference prices then rise again at the end of the season.

The EEC Common Customs Tariff is shown in Table 12.

Table 12

Common Customs Tariff

		per cent of entry price
LETTUCE and Endive		
	1 April - 30 November	15
	1 December - 31 March	13
TOMATOES		
	1 November - 14 May	11
	15 May - 31 October	18
CUCUMBER		
	1 November - 15 May	16
	16 May - 31 October	20
CELERY		
	1 January - 31 December	16

This is not as sophisticated a schedule as the one for reference prices but the Tariff amounts to a high levy on produce exported to the EEC. Exporters have good reason to make every effort to sell at prices above the reference prices, below which there is a countervailing charge, because they receive a good price for their produce, but they cannot escape the tariff.

Accession to the EEC for Spain and Portugal in the fruit and vegetable sector will take place over a transitional period of ten years from the date of entry, 1 January, 1986. For Spain the first phase will last four years while for Portugal it is at present set at five years. However, this first phase may to some extent be regarded as a standstill period. The countervailing charge to Spain and Portugal will fall by small amounts: year one by two per cent, year two by four per cent, year three by six per cent, year four (and, if appropriate, year five for Portugal) by eight per cent. The CCT will be phased out for nonreference priced products progressively to become zero in ten years' time, while that on reference priced products will be reduced by ten per cent in each of the first four years, by 25 per cent in the fifth year and by 15 per cent in the sixth, with five years then seeing a reduction of four per cent in each year so that on the first day of the eleventh year after accession it will be zero. After four years the duty element in the entry price will be diminished by one sixth in each year. The entry price compared with the reference price will become more favourable to Spain and Portugal.

Negotiations with relation to Mediterranean third countries are being conducted at present and the EEC hopes to make a new trade agreement with these countries in the foreseeable future. This agreement will be based on the EEC Foreign Ministers' declaration made last spring, and will give the Commission a mandate for negotiation to achieve a new trade agreement. The Foreign Ministers' declaration reaffirmed the EEC commitment to continue the traditional flow of goods between the EEC and third countries. This commitment clearly states that a new agreement will take into consideration each country's average quantitative export record in the last five years. It is agreed that a review of the new agreement, expected in September 1986, will be made in 1996, when Wpain and Portugal become full members. Mediterranean third countries to be covered by the agreement are Morocco, Algeria, Yugoslavia, Israel, Egypt, Tunisia and Cyprus. The terms will specify and define trade relations as to quotas, supply timetables and customs rates. New products to the negotiations include lettuce and Chinese leaves which will be subject to limited and defined marketing periods and quotas for which no customs will be imposed. For any quotations not within the agreed marketing periods the present custom rates will be applied. It is very probable that Mediterranean third countries will be better off than at present under the new agreement, but at the same time Spain and Portugal are set to be in an unassailable position compared with other Mediterranean countries as they move to complete assimilation within the European Economic Community.

OUTLOOK FOR SALADS

The salad market in Britain has a most promising future. In the case of lettuce, consumption has increased greatly. Summer demand is mainly of interest to home producers but there is evidence of the emergence of a strong winter demand mainly supplied by imports from third countries with warmer climates. The demand is so great as to attract supplies from a number of distant countries because the winter price is so high in accordance with the strong demand. The outlook for lettuce is now given a new boost because of the increased demand since 1982, which has been accompanied by more imports.

Tomatoes have not witnessed the same large shift in recent consumption as lettuce but the last three years have shown a distinct increase. At the same time imports have become a larger part of supply and this trend, so evident since 1980, is likely to continue. Unlike lettuce, imports of tomatoes are dominated by a very few countries: Holland for the home season and the Canaries and Spain catering for the British off-season. While there is little scope for countries to compete with Holland, there is the opportunity for Mediterranean third countries to export some tomatoes in the off-season,of home supply, and many of them have been doing this for some time.

Cucumbers have shown a bigger expansion in consumption than lettuce in recent years and the always significant proportion imported has increased markedly. Consumption, from a low point in 1973, has more than doubled and is now on a new rising trend. Again as with tomatoes a few (in this case two) countries, dominate imports: Holland in summer, and the Canaries in winter. A number of Mediterranean third countries compete with the Canaries in the winter supply and they have scope for more imports in meeting the rising winter demand.

Celery has increased in consumption only slightly over the years. For this commodity home suppliers dominate the market. The home-grown field crop is important and this shifts domestic supply further into the winter than the other salad produce. Imports do not really start till November and they come in strongly until May, when the home-grown crop takes over. In the off-season imports are dominated by Spain and Israel and there is scope at this time for other

Mediterranean third countries. The present trend in rising consumption gives every prospect of an expanding market in the winter period.

Britain is becoming more salad conscious and home and imported supplies are working together to meet the rising market. Spending on salads, apart from their obvious importance to health, also depends on trends in disposable incomes, which are now on a rising trend, above that of past years.

There is ample opportunity for third countries, and Mediterranean third countries in particular, to export salads to Britain in the off-season of home supply. The criteria is quality and freshness. The off-season prices are most attractive to exporters and will repay transport costs. In the home season of supply exporting is a very different matter.

English growers can guarantee quality and freshness, as also can the North European exporters, especially Holland. However, the quality specifications, sometimes over and above those of the EEC, which are demanded by supermarkets for English produce, must also be met by any exporter from whatever source. Consumption is expanding. The demand is not for quantity, though this will of course increase, but for quality, and any country to succeed in the export trade to Britain must be fully aware of this. The cold chain now ensures that the produce also arrives fresh. Freshness and quality are the key to the British market and competition on this basis will be intense. Health as a principle basis of demand, apart from the more sophisticated attributes of salad eating, is now a most important underlying factor in demand. A recent report of the Royal College of Physicians cites Britain as one of the most unhealthy countries in Europe. Exporters of salads can help Britain as well as helping themselves and home producers can benefit from a promising market.

PART III
HISTORICAL PERSPECTIVES

9 Israeli cotton: record yields vs record costs

SAM ROTH

In 1953, from the outset of cotton growing in Israel, severe constraints of land, water and labour meant that maximum profits could be achieved only through the intensive use of these resources together with large capital investments. The constraints on the primary factors of production are demonstrated in the following tables :

Total Agriculture in Israel

Year	Cultivated Land (ha)	Water usage (10^3 m^3)	Total Employment
1949/50	248,000	332,000	N/A
1960/61	407,500	1,000,000	121,000
1970/71	410,000	1,249,000	90,000
1985/86	435,800	1,434,000	87,000

Source : Central Bureau of Statistics.

Cotton Growing in Israel

Year	Area (ha)	Yield (kg lint/ha)	Yield (kg lint/cu m)	Yield (kg/worker)
1963/64	13,000	1,030	0.215	69
1970/71	32,300	1,090	0.245	84
1980/81	57,700	1,350	0.300	120
1985/86	65,000	1,530	0.320	170

Source : Israel Cotton Board.

The absolute price of the inputs tends to rise as various crops compete for them relative to those inputs which are more in abundance. For example, the average cost of water for farming in Israel is about $ 0.11 per cubic meter of water vs. the case for the Californian farmer who has much less of a water problem and pays about 1/10 of this amount.

Under these conditions, optimisation (maximum total profit to the grower) occurs at the point in the production curve where high input costs produce almost maximum yields, rather than a point of minimum cost per pound of cotton. This is in contrast to many other countries where abundant resources take the optimal production point as that at which the production costs per pound are minimal. A comparison of Israeli cotton yields to those of other countries shows that Israel has the highest yield in the world.

Country	Yield, kg lint/ha
Israel	1,550 (not including Pima)
Australia	1,240
China	780
USSR	780
United States	665
Pakistan	415
India	210

5 year averages.

Source : ICAC (Intl. Cotton Advisory Committee).

Some individual farmers in Israel obtain yields of above 2,000 kg/ha. Over the past 22 years (1963 - 1985), average yields have increased by 48.5%, an annual rate of 1.9%, which is more than double the rate of increase in the USA. Higher yields than those of other countries are partially a reflection of economic incentives and partially a question of technological advances.

The economic incentive has further developed the natural tendency of the farmer to see record yields as a proper measure of professional success. The majority of cotton growers in Israel (85%) are cooperative kibbutzim in which the individual's standard of living is determined by the overall economic situation of the community and not necessarily by the profitability of the individual's particular job or branch. The cash flow of the community is centrally managed. Therefore, the individual is often somewhat isolated from short-term economic information regarding optimal yield decisions. Status in the kibbutz and the surrounding communities is often related to yield achievement. A prize is awarded annually in each of the nine growing regions for the highest yield. No such prize exists for the greatest economic efficiency.

All of this has been achieved by intensive research and development particularly in irrigation and fertilisation techniques. Drip irrigation accounted for 35% of the total area in 1985. In 1986 this percentage was estimated to be substantially higher. Total area devoted to cotton dropped from 65,000 ha in 1985 to 46,000 ha in 1986. All of the previously existing drip irrigation systems continue to irrigate cotton. The remaining areas to be watered by sprinkler are relatively small compared to the previous year. Annual expenditures on direct agricultural research pertaining to cotton growing, not including product research by industries (such as irrigation and field equipment manufacturers) approach $2,000,000 (about 1.5% of the total income from cotton-fiber and seeds).

A higher price for cotton means an increased incentive for higher yields. The marginal cost of production should rise to the point equal to the marginal price. From the other direction, a high marginal cost of production demands a high-priced cotton.

The price of one variety of cotton relative to another is related to its physical characteristics which affect its spinability in a textile mill. These include fiber length, colour, trash content, fineness, maturity, uniformity and many other features. Israel grows two varieties of cotton: Acala, a long staple upland cotton, and Pima, an extra-long staple cotton. The Acala variety is priced in the top 15% of the world's cotton. Recent advances in Israeli yields of extra-long staple Pima cotton have allowed that variety to jump from 4% of the total area in 1985 to 28% of that in 1986. This cotton is priced in the top few percent of the world's cotton, about 35% above the Acala variety. It yields in Israel about 15% less than Acala whereas in other countries the gap is much wider (e.g., the USA gap is about 33%). It is also more expensive to grow and gin. Each grower must calculate the most economic combination of varieties according to yields and costs of production.

Over the years, production costs have increased in nominal and real dollar terms. This is partly due to the use of increased amounts of physical inputs and partly to increased costs per unit of input. For example, the cost per unit of the basket of production inputs in cotton in nominal dollars increased by 23% from 1983 to 1986. The increase after the dollar inflation in USA for this period is about 10%. A lack of reliable data prohibits any exact cost comparison of today to the past. The data available seem to indicate that the value of marginal yields is greater than the additional production costs under normal circumstances.

Given these conditions, there exists a significant risk that short- or long-term changes in relative prices may make certain technological advances profitable, despite the fact that they increase yields and in the past produced positive marginal return.

For instance, the use of drip irrigation can increase yields by 250 - 350 kg of fiber per hectare over those achieved with conventional sprinkler systems. Additional costs of production are incurred in the amortisation of the drip system, increased amounts of and the use of more expensive fertilisers. A short-term fall in cotton prices relative to these inputs, similar to that which occurred in 1985 and 1986 raises the break-even point of yield for investing in drip irrigation (the point where the value of the marginal yield resulting from drip irrigation equals the additional cost). A drip irrigation system has a six to seven year life span. Therefore, the farmer makes a decision about such an investment with a long-term perspective. Once the system is purchased there is little room for short-term reaction to relative price changes. The system has become a yearly fixed cost.

On the other hand, Israel suffered a water shortage in 1986 which resulted in about a 10% reduction in water quotas to farmers. This is an example of a change which must be dealt with in short-term as well as long-term decision making. No one knows when the current shortage will end, but it is something that could occur again in the future.

This brings to light Israeli experiments which have been conducted since 1984 on skip row cultivation. Maximum yield per unit of water becomes the rational economic behaviour in the case of a water shortage. It is achieved through extensive use of land (1.5 - 2 times the conventional method). The yield per hectare drops as the return per unit of water increases through more efficient irrigation techniques. The optimal combination of factors of production depends on the shadow prices of water and land. As water becomes more scarce, its shadow price rises, even if the nominal price remains stable.

The following graph is based on a rough analysis of initial field testing of this method. It should be viewed primarily as academic rather than in absolute terms. Further field testing is being carried out and has the potential for later commercial applications.

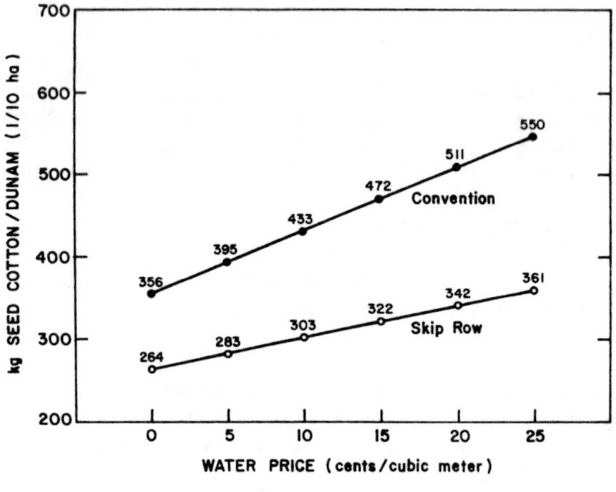

The graph indicates that as the price of water rises and the cost of land and the price of cotton are kept constant, the break-even point for conventional cultivation increases proportionally greater than that of the skip row technique.

Israeli researchers are only beginning to study the potential of skip row techniques in relation to water utilisation. It is too early to conclude if this solution is optimal during times of water shortages. If it is so, it has great advantages in its flexibility. Land preparation before the winter rains is identical for both types of cultivation. The grower needs to decide which technique to use only a short time before planting, thus eliminating the element of risk in predicting months in advance the availability of water. Of course, the optimal situation may be a combination of skip row and conventional cultivation, and linear programming will give us the answer to this problem.

In conclusion, Israeli farmers must reexamine their farming practices with optimal yield in mind rather than maximum yield. If skip row cultivation is proven to be economically viable, it will require a revolutionary change in thinking by Israeli farmers to abandon record yields as a measurement of success.

10 History and analysis of growing tomatoes for processing in Israel

CHANAN GEISENBERG

In the early years of agricultural development in Israel, the processing industry was based on the supply of surplus tomatoes from the fresh tomato market. The quality was not satisfactory, and the quantity flunctuated. The factories could not rely on the supply, and production was therefore inefficient. Efficient production could only be achieved with better fruit quality through planned tomato production, which would ensure a continuous supply. It was found that this aim is basically common to both grower and processor. As a result, cooperative processing plants, which belong to grower organisations, now operate in Israel and carry out about 60% of the processing tomato production.

The main goal of the industry, for processors and growers alike, is a continuous supply of high quality tomatoes. Taking into account climatic limitations, tomatoes in Israel can be supplied to the processors from the end of June to the end of September, i.e., for about 90 days, excluding the Sabbath and holidays (see Figure 1). However, each climatic region has its optimal time of production, and the regions, after several years of trial, have been more or less defined. Nevertheless, adaptation of optimal production to the various areas has already been experimented upon for many years with regard to varieties, irrigation and fertilisation, as these factors interact with soil and climate. This research will be continued.

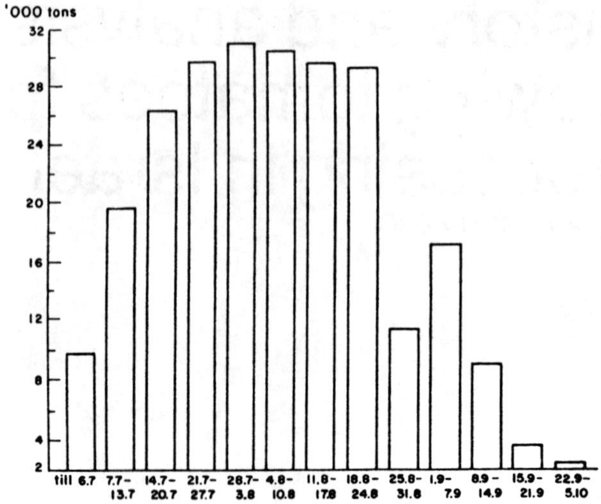

Figure 1 Weekly supply of tomatoes to factories, Israel, 1985

The production of tomatoes for processing in suitable areas might also be limited by available labour, mainly for harvest, and by available water for summer or early autumn production. Alternative crops might be preferred by potential growers for economic or other reasons.

Low temperature in early spring and decreasing yields during the summer high temperature period together with increased production costs for irrigation and plant protection, explain the relatively low supply of tomatoes in June and from the last week in August onwards, as shown in the Figure 1. In order to encourage production during these marginal periods, a premium is paid by the processors to the growers for early and late supply of tomatoes. Also, a graduated premium is paid on each batch of fruit supplied if the Total Soluble Solids (TSS) content is above a certain level, while below this level, a lower price is paid.

Advantage is taken of the regional climatic conditions, thus warm areas are utilised for early production (Jordan Valley) and the mountain areas (north and north-east in the Golan Heights) for late summer production.

The research on tomatoes for industry was led during the last 18 years by the late Professor Jehoshua Rudich of the Faculty of Agriculture of the Hebrew University of Jerusalem.

VARIETIES

Open-pollinated and hybrid varieties are bred for earliness, resistance to high temperatures, high TSS content, peelability, good internal colour and taste, resistance to soil-borne, leaf, stem and fruit diseases and suitability for mechanical harvesting and bulk transportation (Miltau, 1985).

Results of the breeding programmes, which are applied in commercial fields, are presented in Table 1.

Table 1

New tomato varieties for processing

Marketable yield of four varieties and TSS measured in % Brix. Average of ten trials, planted between 15 March and 10 May, 1985. Average of six replicates in each trial, 19.2 sq m each plot (Rudich, 1986).

Variety	*Marketable fruit (kg/sq m)*	*% Brix*
M82-1-8	10.8	4.7
Hofit	10.6	5.4
Holit	11.1	4.8
S80-285	11.9	4.6

M82-1-8 is the standard variety grown in 80% of the fields in recent years. It is resistant to VF (*Verticillium* and *Fusarium* wilt).

Hofit has a 15% higher TSS than M82-1-8 and is resistant to *Fusarium* wilt race 1 and race 2 and to *Verticillium* wilt. It has very good colour and is suitable for all product usage.

Holit is similar in yield and shape to M82-1-8, but has better internal colour and therefore is better suited for peeling. It is resistant to VF_1 F_2 (*Verticillium* and *Fusarium* wilt, race 1 + race 2).

S80-285 is adapted to higher temperatures and achieves relatively higher yields. It is resistant to VF.

New hybrid varieties will probably be approved for use in the coming years. Some of them were already tested last season in commercial quantities in a few processing plants, with good results. They will be introduced if they show economic advantages (yield and quality).

PLANT DISTANCE

The envisaged commercial use of hybrid varieties, with their relatively high seed price, induced us to reconsider the commonly accepted planting distance. Another aim of this investigation, which has not been concluded so far, is to reduce labour-input for thinning out to single plants in directly seeded fields by seeding to a final stand. The accepted plant population is 50,000 - 60,000 plants per hectare.

The results have shown that with trickle irrigation, varieties of good vine-storage, such as M82-1-8, exhibit no significant difference (p = 0.01) in marketable fruit yield with population of 25,000 - 100,000 plants per hectare (see Table 2).

Table 2

Plant population trial with var. M82-1-8

Planted 20 May, harvested 9 September 1985. Trickle irrigated. Average of six replicates, 19.2 m each plot (Rudich, 1985).

Plant/sq m	*Marketable yield (kg/sq m)*
20	8.57b
10	11.25a
5	11.54a
2.5	11.15a

S.E. 0.263
Significance $p = 0.01$.

SEED TREATMENT

Seed treatment with $KNO_3 + K_3PO_4$ is widely practiced for early seeding under suboptimal soil temperature conditions, for enhanced germination, better seedling development and early yield.

HIGH PHOSPHORUS FERTIGATION

Liquid fertiliser 8:24:0, sprayed on the seeds in the seed bed, is considered helpful for early emergence and development of the plants under suboptimal temperature conditions. But it was found safe and profitable in some kinds of soils and harmful in others. The combination of seed treatment with $KNO_3 + K_3PO_4$ and 8:24:0 fertigation is practiced in many places in early spring.

ENHANCEMENT OF RIPENING PROCESS

The use of ethephon was found practical in autumn, when temperatures are relatively low and heat-units are a limiting factor for normal concentrated ripening in mountain areas (e.g., Golan Heights).

WATER-STRESS AND SALINITY FOR IMPROVEMENT OF TSS-LEVEL AND OTHER QUALITY FACTORS

Water-stress during fruit development and the fruit ripening process was found to have a marked influence on fruit quality of many varieties with regard to TSS content of the fruit (Figures 2 and 3). Also better internal colour was observed under these conditions. It was recommended that the grower applied water stress during the fruit-ripening process as this procedure only slightly influenced the yield level, whereas the income per planted area was greater owing to the better price paid for higher TSS content of the fruit. On the other hand, reduced expense for water, harvest and transportation made the procedure economically preferable to higher yield as a result of higher irrigation rate.

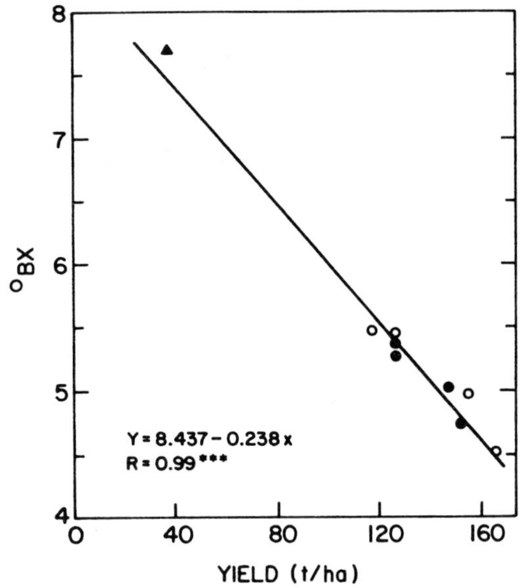

Figure 2 Relation between Total Soluble Solids (% Brix) and yield of fruit supplied to industry.
Key: empty circle - drip; filled circle - sprinkler; filled triangle - rainfed

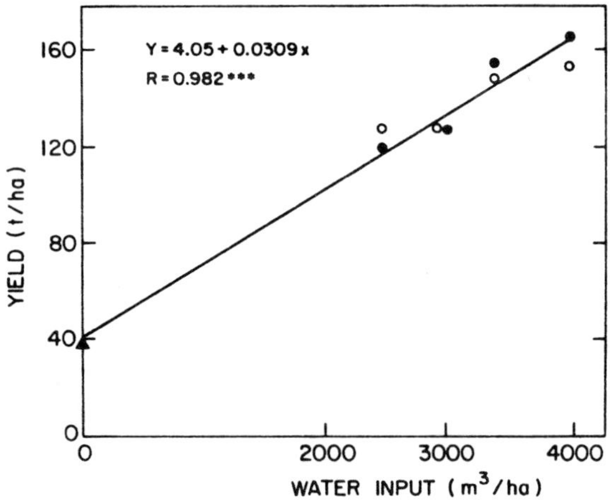

Figure 3 Relation between the amount of irrigation water and yield supplied to industry, 1983

A similar effect was shown by the use of saline water up to 6 millimos, provided either transplants were used or fresh water was used in a direct-seeded field for the first three weeks until plant establishment (Orly, 1984).

REFERENCES

Atherton, J.G. and the late J. Rudich (eds.), (1986), *The Tomato Crop*, Chapman and Hall, London, New York.

Miltau, O., (1985), *Field Observations and Investigation of Performance of New Processing-Tomato-Varieties in Tomato-Processing-Factories*, The Hebrew University, Jerusalem. Faculty of Agriculture, Rehovot (in Hebrew).

Orly, C., (1984), *The Effect of Potassium Chloride Salinity on Yield and Quality of Processing Tomatoes*, The Hebrew University, Jerusalem. Faculty of Agriculture, Rehovot (in Hebrew with English summary).

Rudich, J. and Geisenberg, C., (1984), *Research in Tomatoes for Processing*, The Hebrew University, Jerusalem. Faculty of Agriculture, Rehovot (in Hebrew).

PART IV
OPTIMAL USE OF AGRICULTURAL INPUTS

11 Physical models for evaluating the influence of various management practices under optimal yield management

PHILIP V. RASMUSSEN AND JOHN R. HANKS

INTRODUCTION

Modelling or systems analysis was first described in detail three hundred years ago by Newton (1687). Newton reasoned that by creating mathematically correct models, we could describe the future motions and positions of the planets with great accuracy, without the necessity of actually taking real-time measurements. Hence, if a system were properly understood, mathematical models could be used to predict the outcome of a given set of circumstances on paper or blackboard without the expense or danger of actually creating those circumstances. As man's knowledge of the physical and biological universe about him has increased, interest in such systems analysis as developed by Newton has also increased. An additional stimulus towards management of models has been the development of personal microcomputers with large cache memories that can be used to simulate management systems. Today we are entering a new era in agricultural management in which the outcome of management options can be simulated prior to the actual implementation of those options. Several options can be quickly simulated on a personal microcomputer and the most profitable option chosen for actual use.

Numerous researchers have described the possible use of systems modelling principles in agricultural systems. Norman (1981), Sakamoto and LeDuc (1981) and Sampson (1984) discuss many of these uses, among which are the following :

1. Evaluation of management of scarce resources by evaluating various options through nondestructive (computer simulated) testing.

2. Evaluation of management options that are too costly or too dangerous to actually perform, such as the influence of a nuclear winter on worldwide wheat production.
3. Evaluation of management options that have prohibitively large time-scales for actual testing, such as evaluating one hundred years of irrigation with saline water at a given site.

Many other uses for systems modelling can be listed. However, it is sufficient to understand the possible uses of systems models for a farm manager who could use them to effectively evaluate the profitability and crop yield potential of his management options *before* implementing them.

At the present time, the *key problem* lies in delivering the models to the end-user, a farm or ranch manager, in a form that he can use them effectively with little or no specialised training. Numerous models for various crops, irrigation schemes and soil phenomena exist, but few are in a form that is usable by any but highly trained scientists or technicians. Our approach has been to take the principles of several complex, specialised models, that up till now were only used on large mainframe computers, and convert them for use on personal microcomputers. Several on-farm users have utilised these models for management decisions by implementing them on home computers. However, we are currently amending these models with artificial intelligence shells to give them a more conversational nature. Physical simulation models can be used by most managers to develop optimal yield strategies. Models are also useful for technically trained people to use for simulation of many situations and to produce production functions.

MODEL DESCRIPTION

The models that we have currently adapted to use on microcomputers have been described thoroughly in other publications. PLANTGRO is a BASIC or FORTRAN-coded model of crop water use and crop yield that has been tested for corn (Hanks, 1974), spring wheat (Rasmussen and Hanks, 1978), alfalfa (Retta and Hanks, 1980a and 1980b) and many other crops (Hanks, 1983). Its approach is to split the potential evapotranspiration (PET) value into potential soil evaporation ($E\text{p}$) and potential plant transpiration ($T\text{p}$) (Childs and Hanks, 1975; Hanks and Puckridge, 1980; Rasmussen and Kanemasu, 1979). Relative yield is assumed to be directly related to relative transpiration ($Y/Y\text{p} = T/T\text{p}$). Actual soil evaporation (E_A) is assumed to be equal to $E\text{p}$ divided by the square root of the time (days) since last wetting (Ritchie, 1972). Actual plant transpiration (T_A) is calculated from subsurface layer water holding capacity data that assume T_A to equal T_p until a critical amount of water holding capacity is lost (typically 50% in vegetative phases and 35% in critical grain filling stages or with succulent crops such as tomatoes) after which T_A then decreases linearly to zero as the available water goes to zero. Yield is calculated by growth stage using the assumption that T_A/T_p can be related to the relative potential yield (Y_A/Y_p) (see Figure 1).

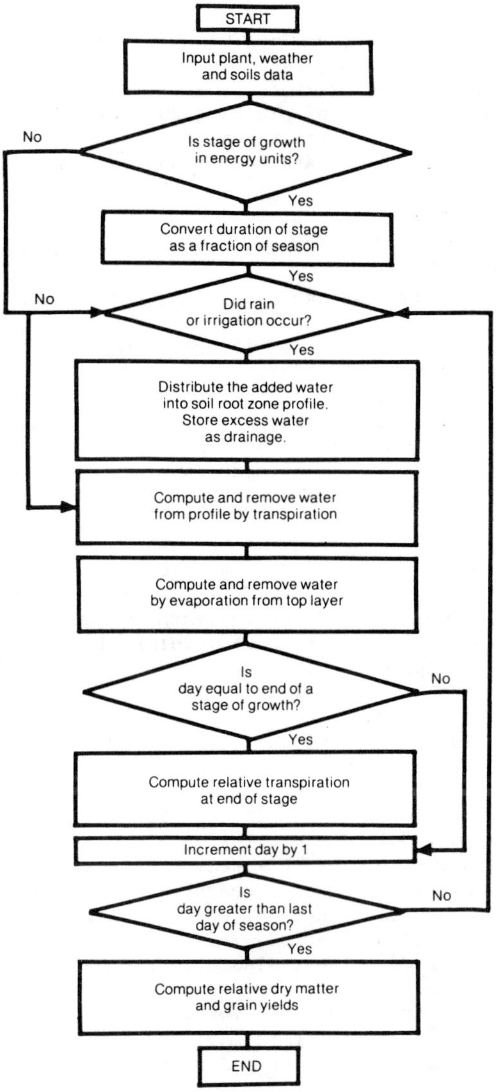

Figure 1 Simplified flowchart of PLANTGRO model

The inputs required are :

1. Soil information on water holding capacity and initial water content of each layer.
2. Climatic information on PET and day and amount of rainfall.
3. Crop information on dates of planting and harvest, depth of rooting vs. time and critical growth stages.
4. Management information on data and amount of irrigation.

The outputs available are :

1. Cumulative A_A, T_A, E_A. T_A, ET and drainage below the root zone as a function of time.
2. Relative water content of any layer as a function of time.

The output data may not all be of interest and can be suppressed if desired. The detailed output data are useful for instruction purposes since the date at which plant water stress occurs can be determined. This simple model allows reasonable answers to be given to such questions as:What effect on yield would I get if I irrigated with 10 cm of irrigation water on 4 July? The model also makes the user aware of all the other information needed to answer such questions.

This same approach was combined with a complex biometeorological time scale, to drive an ET_A/ET_P described by Kanemasu, *et al.* (1977 and 1978). It was later improved to simulate yield/water-use entitled WHEAT-ET coded in IBM-G-level FORTRAN. A comprehensive review of both of the above models and predicted vs. observed yields of each is given by Hanks and Rasmussen (1985) and Hanks (1983).

The PLANTGRO model has been extended (Retta and Hanks, 1980a) into an irrigation simulation management tool entitled IRRIGATE-MATE coded in the relevant BASIC language of several microcomputers, including Apple, IBM-PC and ATARI-520ST. A user can sit at the computer and simulate many years of production in a few minutes with options that range from a complete weekly soil moisture monitoring and a start-of-season soil moisture survey, to a mere daily advisement of weather and rainfall. The user is graphically shown soil, weather, irrigation and yield conditions through the game and is given a calculated net profit at the end of a season based on irrigation, fertiliser, soil testing and water monitoring costs incurred. Also included in the calculation are penalties for yield decrease due to crop stress and loss of fertilisers if leaching occurred beyond the root zone. This has proven to be a highly effective tool in teaching farmers and students the relationships of soil moisture status, irrigation management, and optimal economic crop yields.

Rasmussen *et al.* (1986) have developed a similar model for IBM personal computers entitled CONTIL that simulates yield vs. evapotranspiration relationships for high-mulch conservation tillage systems in spring and winter wheat. An energy partitioning function further reduces E_A from E_P, according to the amount of surface stubble (residue) that is left by such practices as no-tillage or minimum-tillage seeding system. This approach has also been used for partitioning the energy of evapotranspiration in the high surface-residue conditions of grazed rangeland in a model entitled PARANET (see Figure 2) for the IBM-PC (Belliston, 1986).

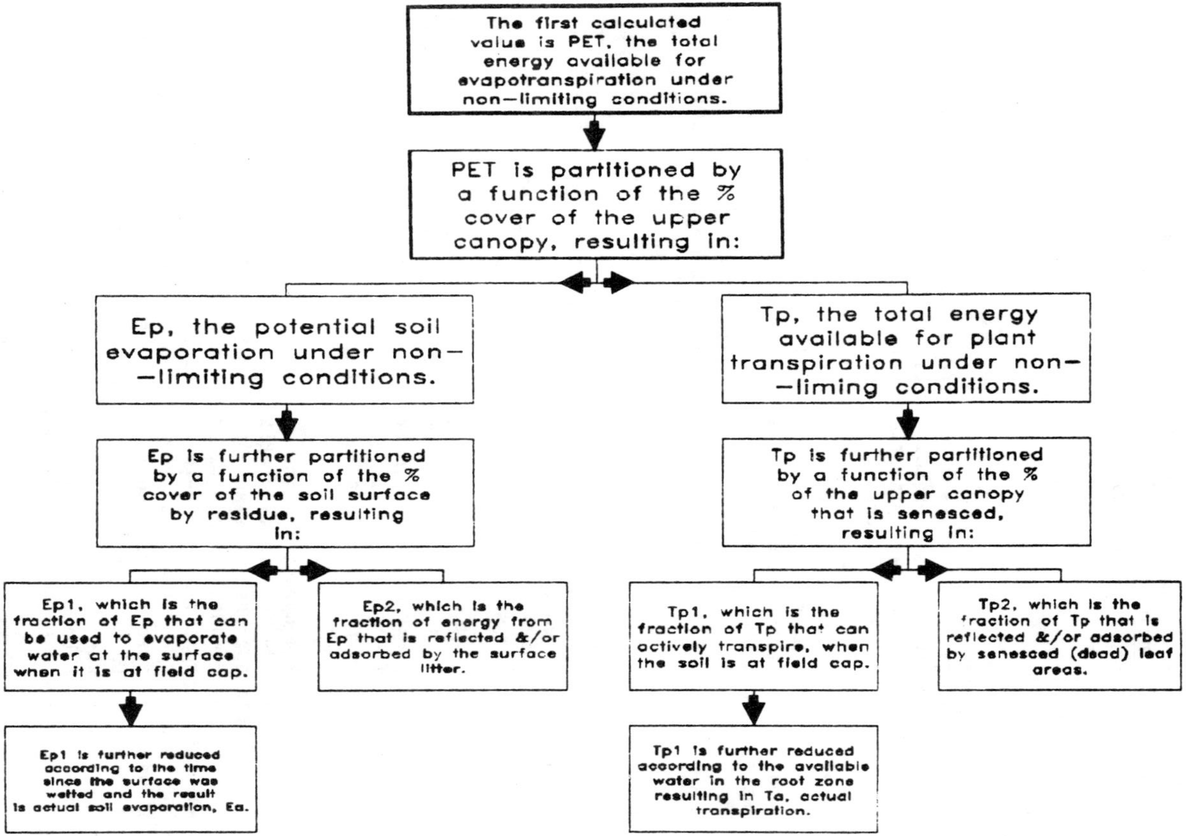

Figure 2 Partitioning of PET by CONTIL and PARANET

These rather simple models work well if all water movement is downward and if salinity is not a problem. If there is a water table close to the surface so that water can move up as well as down, a more sophisticated model is called for. Similarly, if salinity is a problem it is not sufficient to consider water content only - water and osmotic potential must also be accounted for.

The model SOWATSAL (Childs and Hanks, 1975; Hanks and Hill, 1980) has also been used to include the influence of soil salinity and/or the presence of a water table on yield prediction (Hanks and Andersen, 1981).

DISCUSSION

The basic mechanisms of the models described here are similar. Actual evaporation and transpiration is calculated by partitioning evaporative energy and evaluating the stored soil water and soil water and soil surface conditions. Yield is estimated by transpiration/yield relationships unique to each crop. These models have then to be extended into management models on personal microcomputers that predict yield and possible profit/loss relationships vs. irrigation or surface mulch management. These models (IRRIGATE-MATE, CONTIL and PARANET) have been used in educational classes, in extension workshops with landowners and operators and by farm and ranch managers to predict possible consequences of given management practices *prior to implementation*. Thus, optimal economic yields have been obtained by previously testing possible outcomes of management methods and choosing the optimal course.

The more complicated model SOWATSAL has usually been used in research situations. The input data requirements are somewhat formidable and the computer time required is many times that of the simpler models. Nevertheless, if the salinity or water-table problem is of concern, it has given reasonable simulations when simple models do not predict well.

The most obvious problem in taking a complex systems analysis approach with complex physical models is in extending their use to the untrained manager. Several hours of intensive training is needed prior to independent use of PARANET, CONTIL, IRRIGATE-MATE or PLANTGRO on a microcomputer. We are currently amending these models with artificial intelligence modules so that :

1. Natural language associative memory interpreters can be used. Thus if a person responds What is my profit? to a query, the computer gives several possible options. Once an option is matched with a natural query, they are permanently *associated* in computer memory. Thereafter, when What is my profit? is typed-in, the computer immediately responds with the proper answer that was previously *associated* with that query.
2. Bit-image pattern recognition is used so that a mistyped response YSE is taken as YES because their bit image is similar. This allows the computer to tolerate a certain amount of error.
3. Cognitive data base searching is possible so that conditions of a given area will not have to be tediously typed-in at each session.

These additions to the models will allow their use by the broadest possible public. In addition, the associative memory option will allow more rapid adaption to foreign language users (see Figure 3).

Models cannot evaluate all possible options, nor be totally accurate under every condition and climate. However, we feel confident that the aforementioned models can be used to evaluate irrigation and surface residue management with reasonable accuracy and are a valuable aid to train managers in optimal management techniques.

Limitations of evapotranspiration-based models

It is *crucial* that the user be ever mindful of the limitations of a given model. Unless he or she is extremely cautious when entering initial conditions and the subsequent viewing of the output, an undeserved trust for the black box may develop. Adequate crop yield prediction is possible as long as a *proper* Y_A is measured for a given site and the *most limiting factor is water stress*. When simulating yields under differential irrigation schemes, adequate yield prediction is possible with such models if some stress occurs. However, if some other factor is limiting, this approach cannot be used with confidence.

An example of the possible problems that limiting conditions (other than water stress) can produce is illustrated in Figures 4 and 5.

Figure 4 shows the differential yields that result from differential water management schemes at Riley County, (Eastern) Kansas as modelled by Rasmussen and Kanemasu (1979). However, in Figure 5, wheat yields at Ellsworth and Finney counties (Western Kansas) show a divergence in yields that is not explained by relative transpiration ratios. The R^2 of observed vs. predicted yield values drops from 0.85 in Riley Country to 0.55 in Ellsworth and Finney Counties. Obviously, something other than water stress is causing the yield variation. The management models, described herein, assume that the *only* variable between comparable alternatives is one that will effect the T/T_p ratio. Any other variables such as fertilisers, variety (cultivar) or hybrid seed type used, pests and pesticides, or disease and disease control, *cannot* be accounted for. Therefore, a crucial limitation of these models is the selection of a *proper* potential yield value according to :

1. Variety or hybrid used.
2. Climatic temperature and precipitation (year).
3. Soil type fertility level.
4. Soil fertility inputs.
5. Soil tillage and management patterns used.
6. Inherent pest, disease and other limitations of yield.
7. Disease and pest control measures used.
8. Chemical growth regulators used.
9. Any other management input used.

If a proper reference yield is not chosen, or if variables other than those that directly affect plant and soil water status are introduced, these models become useless. In addition, a sensitivity analysis of these evapotranspiration-based models shows that variations in the available water holding capacity (AWC) of a test site soil and the amount of water available at the time of seeding (BGSM) will substantially alter the predicted yields under a certain climatic or irrigation regime.

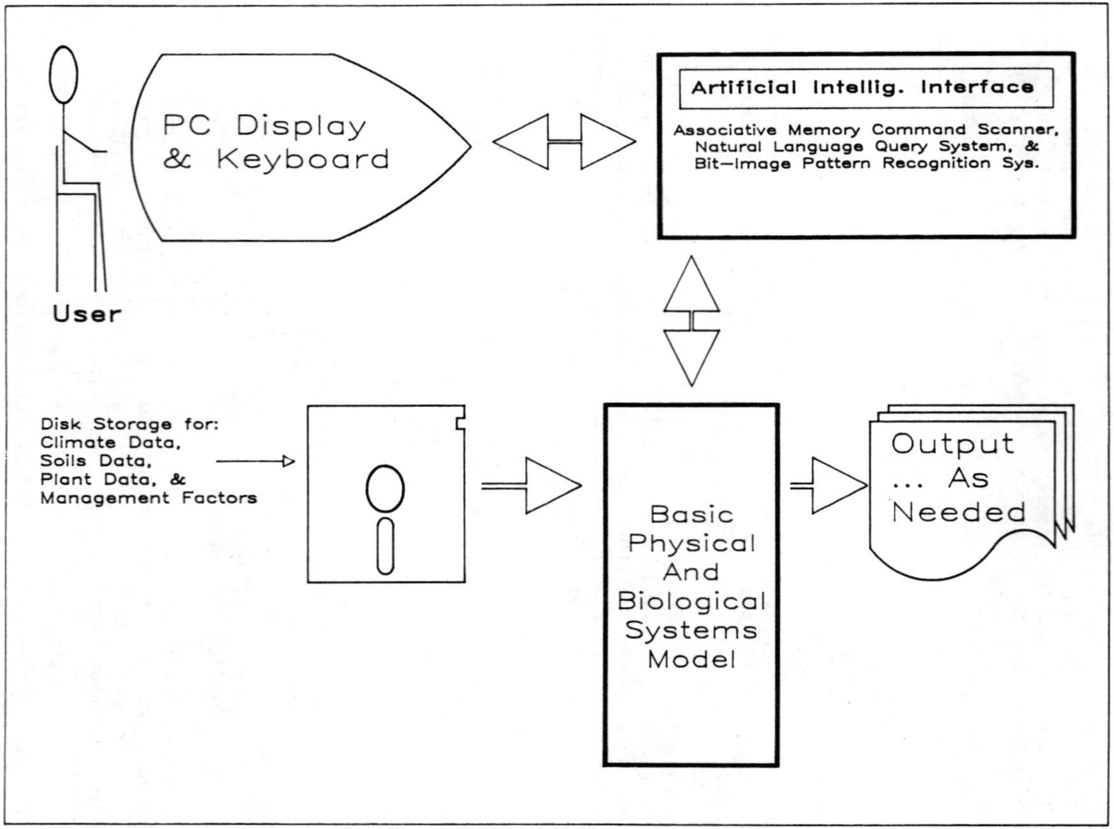

Figure 3 Artificial intelligence to bridge the model/user gap

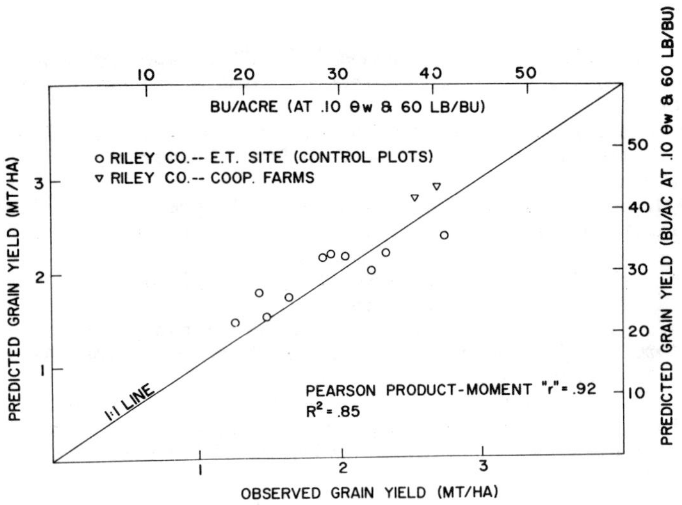

Figure 4 Observed vs. predicted yields where water is limiting

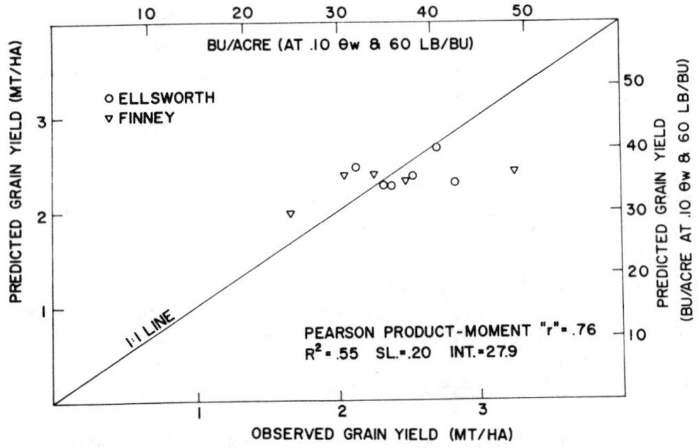

Figure 5 Observed vs. predicted yields where water is not limiting

REFERENCES

Belliston, J., (1986), *A Computer Model for Prediction of Soil Water Status and Drought in Utah*, (MS Thesis), Utah State University, Logan, Utah.

Belliston, J., Gordon, G., Rasmussen, V.P. and Hanks, R.J., (1985), *PARANET: Users Manual and Report*, Utah State University, Logan, Utah, 24 pp. (Revised).

Childs, S.W. and Hanks, R.J., (1975), 'Model to predict the effect of soil salinity on crop growth', *Soil Sci. Soc. Am. Proc.*, Vol. 39, 617-622.

Hanks, R.J., (1974), 'Model for predicting crop yield as influenced by water', *Agron J.*, Vol. 66, 660-665.

Hanks, R.J., (1983), 'Yield and water-use relationships: an overview', in *Limitations of Efficient Water Use in Crop Production*, American Society of Agronomy, Madison, Wisconsin, 393-411.

Hanks, R.J. and Andersen, J.C., (1981), 'Physical and economic evaluation of irrigation return flow and salinity on a farm', in Yaron, D. (ed.), *Salinity in Irrigation and Water Resources*, Marcel Dekker, New York.

Hanks, R.J. and Hill, R.W., (1980), *Modeling Crop Responses to Irrigation in Relation to Soils, Climate and Salinity*, Irrigation Information Center, Bet Dagan, Israel, 66 pp.

Hanks, R.J. and Puckeridge, D.W., (1980), 'Prediction on the influence of water, sowing date, and planting density on dry matter production of wheat', *Aust. J. Agric. Res.*, Vol. 31, 1-11.

Hanks, R.J. and Rasmussen, V.P., (1985), *IRRIGATE MATE: Student Manual*, Dept of Soil and Biomet, Utah State University, Logan, Utah 84322-4840.

Kanemasu, E.T., Heilman, J.L., Bagley, J.O. and Powers, W.L., (1979), 'Using LANDSAT data to estimate evapotranspiration of winter wheat', *Environ. Manag.*, Vol. 1, 515-520.

Kanemasu, E.T., Rasmussen, V.P. and Bagley, J., (1978), *Estimating water requirements for Corn with a 'Pocket' Calculator*, Kansas Agric. Exp. Sta., Bulletin 615, Manhattan, Kansas.

Newton, (1687), *The Principia : The Mathematical Principles of Natural Philosophy*, Translated into English by Motte, A., Printed by Motte, B., London, England, 1972.

Norman, J.M., (1981), 'Data requirements for generalized physiological models', in Weiss, A. (ed.), *Computer Techniques and Meteorological Data Applied to Problems of Agriculture and Forestry: A Workshop*.

Rasmussen, V.P., Belliston, J., Cardon, G. and Hanks, R.J., (1986), *CONTIL: Users Manual and Report*, Utah State University, Logan, Utah (Revised).

Rasmussen, V.P. and Kanemasu, E.T., (1979), 'Modelling winter wheat yields as affected by water relations and growth regulants', *Proc. 14th Conf. Agric. Forest Meteorol. and 4th Conf. on Biomet.*, American Meteorol. Soc., Boston, Mass.

Retta, A. and Hanks, R.J., (1980a), *Manual for Using Model PLANTGRO*, Utah Atric. Exp. Sta., Research Report No. 48, Logan, Utah.

Retta, A. and Hanks, R.J., (1980b), 'Corn and alfalfa production as limited by irrigation', *Irrig. Sci.*, Vol. 1, 135-147.

Ritchie, J.T., (1972), 'Model for predicting evaporation from a row crop with incomplete cover', *Water Resources Res.*, Vol. 8, 1204-1213.

Sakamoto, C. and LeDuc, S., (1981), 'Sense and nonsense', in Weiss, A. (ed.), *Computer Techniques and Meteorological Data Applied to Problems of Agriculture and Forestry: A Workshop.*
Sampson, J.R., (1984), *Biological Information Processing: Current Theory and Computer Simulation*, John Wiley & Sons, New York.

12 Water stress, radiation stress and cotton productivity

CHAIM JOGEV

INTRODUCTION

Low water potential of the plant or low radiation flux densities are defined here as water or radiation stresses, respectively. This is mainly because most of the physiological reactions, such as photosynthesis, that increase the performance and productivity of the plant and many aspects of growth and development are greatly reduced by these stresses.

The reactions of plants to low water potential are well known (Bradford and Hsiao, 1982). Irrigation techniques developed in recent years are able to reduce water stress considerably. Nevertheless, many irrigation practices do not increase yield since high water potential usually enhances dry matter production without guaranteeing yield increase (Fisher and Turner, 1978).

Plant responses to low radiation flux density are also well documented (Bjorkman, 1981); research deals mainly with performance of single leaves or with the ecological aspect of plants growing in shaded places. Plants react not only to total radiation flux densities, but also the amount and ratio of certain wavelength radiation is of utmost importance (Smith, 1982).

Leaves which supply most of the photosynthate are defined as source, while organs such as roots or generative organs, that accumulate material, are defined as sinks. Despite the importance of photosynthetic efficiency, modern cultivars are often no more photosynthetically efficient than their wild ancestors. The improvement in yield came about through changes in plant structure and development or stronger sinks, such as enhanced kernel size and number in wheat and corn or fiber in cotton (Evans, 1975). However, what makes a sink more active is still not very well understood (Wardlaw, 1980). Although this is a genetic property, water nutrition or radiation stresses also influence sink strength. While water or nutrition potential do not differ greatly among different organs, radiation flux densities change significantly in the crop canopy.

A number of experiments were conducted to evaluate the effects of water or radiation stresses under field conditions on cotton yield and quality and water use efficiency.

MATERIALS AND METHODS

Cotton was grown between 1978 and 1985 in Mevo-Hama in the Golan Heights and in Acre in the coastal area. The soils in both locations are similar, containing up to 70% clay with a high water holding capacity. Experimental plots were drip irrigated. Irrigation or radiation experiments had six replications with plot sizes of 8•30 m. Translocation experiments with ^{14}C, intended to detect photosynthate transport from sink to source, usually had 12 to 14 replications. Radioactive marking (cpm per mg dry matter) is defined as specific activity. The specific activity of one organ (leaf bud or boll) divided by the average specific activity of all organs of a branch or plant is defined as the relative specific activity (RSA), which is a measure of the relative sink activity of an organ. If RSA is greater or smaller than one, than the organ has a higher or lower sink activity than the average.

In the first experiment plastic-lined beds, 30 cm deep and 50 cm wide filled with the original soil, with distances of 50, 100 or 200 cm between beds, provided equal soil volume and water and nutrient supply but different radiation fluxes. In the second experiment single plants were grown at a distance apart of 1•2 m. All branches were removed, except two or four sympodial-fruit branches, to produce plants whose leaf, buds or bolls were without water nutrition or radiation stresses. In the third experiment plants were pruned by only two branches, so that one branch was always in full sunshine the other in plant shade or under shade produced by plastic net and paper; in the control treatment neither of the branches was in shade. The total amounts of radiation reaching the branches were similar in plant or artificial shade, but the light quality (wavelength) differed.

To produce different radiation regimes in a cotton crop, water potential was lowered when the growth rate of the treatment was higher than planned. Otherwise water potential was kept high in all treatments.

In the fourth experiment cotton was irrigated or left unirrigated for three periods:

1. From flower bud appearance to beginning of flowering.
2. First month of boll production.
3. From one month after flower appearance till the first bolls mature.

RESULTS

Irrigation was 50% of the potential transpiration in the first period and 80-90% in the subsequent periods. (To maintain high fertility, all treatments received 8 kg nitrogen fertiliser with 25 mm irrigation water at the beginning of every period). Rain provided the soil with 500-600 mm water during the winter (in Israel rains stop in spring). Night temperatures and relative humidity are lower and evaporation is about 10-15% higher in Mevo-Hama than in Acre. The soil moisture in the first period, while plants were relatively small, was 70-80% water content in the upper 30 cm in both fields. In Acre the water potential of the leaf during the first period decreased from -1.1 to -1.5 MPa (1MPa = 10 bars ∿ 10 atm), but dropped from -1.6 to -2.2 MPa in Mevo Hama. At the end of the second period, leaf water potential reached -1.8 MPa at Acre and -2.8 MPa at Mevo-Hama in the nonirrigated treatments. Figure 1 shows the plant height for the different treatments.

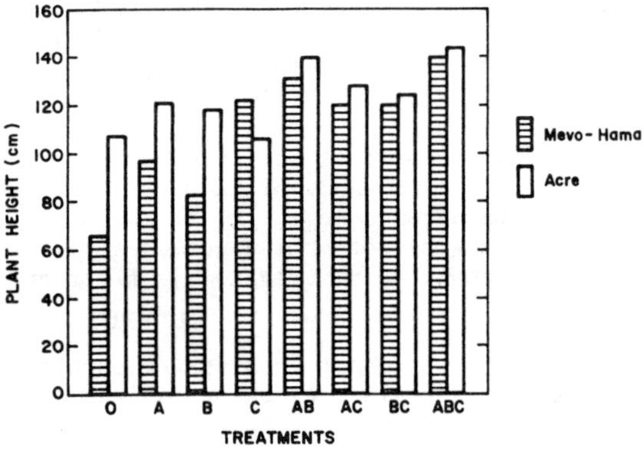

Figure 1 Plant height for treatments irrigated or nonirrigated in three growing periods in Mevo-Hama or in Acre. First period (A) from flower bud appearance to beginning of anthesis; second period (B) main flowering season; third period (C) late boll production. (Example AC = irrigated in first and third period only)

As the water potential in Acre was usually found to be higher, the effect of irrigation on growth was greater in Mevo-Hama. In Mevo-Hama irrigation was necessary in all three periods to produce plants of similar size to those grown in Acre.

The response of cotton yield to irrigation is shown in Figure 2. In Acre yield increased only when irrigation was applied in the third period. In Mevo-Hama irrigating in every period increased yield, but in contrast to Acre, irrigation in the third period only, gave a relatively small response. Irrigation in all periods was necessary to achieve maximum yields in Mevo-Hama. The different responses to low water potential in Mevo-Hama are a result of the relative low humidity in the more inland regions of Israel. Furthermore, low night air temperatures and soil temperatures in spring decrease plant growth, in general, and root growth, in particular. As a result of this, the water potential of

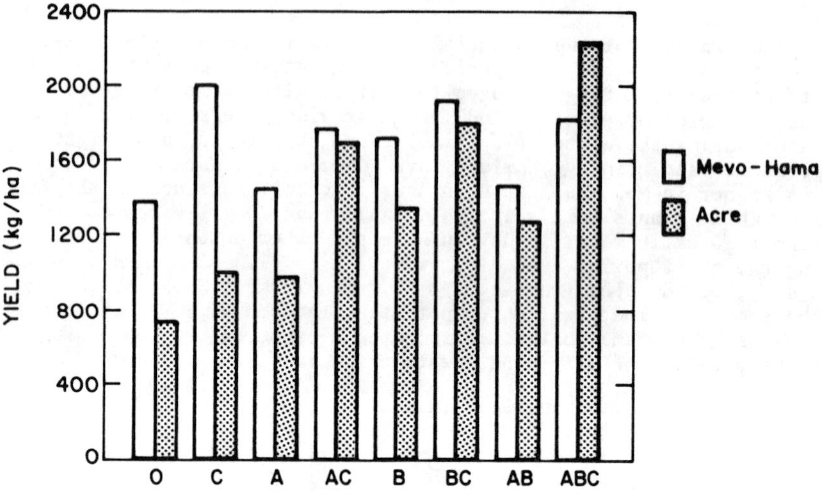

Figure 2 Cotton yield for treatments irrigated or nonirrigated in three growing periods in Mevo-Hama or in Acre. First period (A) from flower bud appearance to beginning of anthesis; second period (B) main flowering season; third period (C) late boll production. (Example AC = irrigated in first and third period only)

the plant is reduced and roots do not grow fast enough to reach the water stored in the soil. Later in the season, as the water demand of the plant increases, the soil's water reserves are depleted which lowers the plant's water potential.

Irrigation experiments during 1978 - 1985 in different regions of Israel, in which irrigation was applied according to the water potential of the plant, gave high yields when the water potential of the plant was kept at -1.5 ± 0.1 MPa throughout the season. If the water potential was relatively high (-1.2 to -1.6 MPa) before the first flower appeared, reducing the water potential at that stage of -1.8 or -2.0 MPa often resulted in similar or even better yields and a much higher water use efficiency.

Monitoring stomata and mseophyll conductance, photosynthesis, nitrate reductase, growth rate and the number of flower buds produced, showed that all were in positive correlation with high water potential. Since most of these experiments were also combined with high fertiliser addition, we looked for some other stress responsible for lowering yields and water use efficiency.

Low radiation flux density reduces photosynthesis (Bjorkman, 1981), cell and plant structure and enzyme synthesis or activity (Nobel and Longstreth, 1981). A single layer of leaves transmitted only 10% of the sun's radiation. In many areas in a fast growing crop, radiation flux densities are extremely low creating radiation stress in many plant parts.

The following experiments investigated the influence of radiation stress is a cotton plant or crop.

Growing cotton in plastic-lined beds produced different radiation flux densities, while keeping the soil volume, water and nutrition supply equal. As can be seen from Table 1, cotton yield reduction due to radiation stress is very severe. As opposed to water stress, radiation stress did not reduce total dry matter yield, but even increased it significantly per square meter.

Table 1

Plant height, dry weight of branches or cotton, ration of cotton/ branches (weight), earliness (first harvest/total yield). Cotton grown in plastic-lined beds with different row distances in Mevo-Hama in 1982

	Distance between rows		
	0.5 m	*1.0 m*	*2.0 m*
Plant height (cm)	173 a	145 b	120 c
Cotton first harvest (g/m)	91.3 c	435.6 b	828.4 a
Cotton second harvest (g/m)	90.6 b	94.3 b	231.1 a
Total cotton yield (g/m)	182.0 c	529.0 b	1,059.5 a
Total cotton yield (g/m^2)	364.0 c	529.0 b	529.7 b
Cotton per boll 1st harvest (g)	6.3 b	7.0 ab	7.5 b
Cotton per boll 2nd harvest (g)	6.3	6.0	6.8
Earliness (%)	50.2 b	82.2 a	78.2 a
Branches weight (g/m)	565	586	719
Branches weight (g/m^2)	1,130 a	586 b	359 c
Total dry weight (g/m^2)	1,494 a	1,116 b	889 c
Ratio cotton/branches (harvest index)	0.32 c	0.90 b	1.47 a

Values followed by the same letter within each row are not significantly different according to LSD 0.05.

If the total dry matter produced is not decreased, then the problem must be in the translocation ability from source to sink, or in sink activity and strength. Figure 3 shows the distribution of photosynthate in a cotton branch when ^{14}C was given to the flower leaf which served as the source leaf. The weight of leaves, bolls and buds decreased apically from the node nearest to the monopodical branch (boll 1) to bud 3 and the apex. A single leaf and boll or bud were allowed to develop on every node. Reducing water, nutrition or radiation stress in these pruned plant branches increased leave and boll size significantly. When supplying ^{14}C to the flower leaf (Figure 3), although the source leaf position is centrally situated, the majority of the photosynthate goes to the cotton bolls (except 39.9% which is left in the leaf 48 h after supplying ^{14}C). With regard to sink activity (Figure 4) the flower adjacent to source leaf tripled the photosynthate translocated per dry weight, while the bud sink activity was lower than that of bolls.

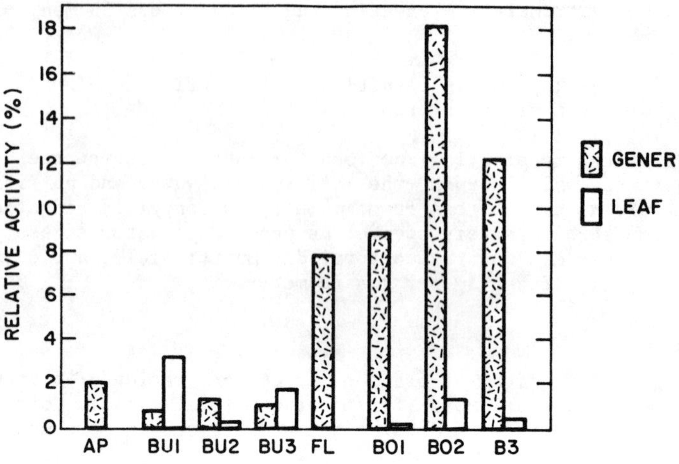

Figure 3 Translocation of ^{14}C from flower leaf to apex, buds, flower and bolls or nearby leaves in a fruiting branch of a cotton plant pruned to four fruit branches. Percent of isotopic marking found 48 h after application. (39.9% were found in the flower leaf, the source leaf). AP = Apex, BU1-3 = bud 1-3, FL = flower leaf, BO1-3 = boll 1-3, GENER = generative organ

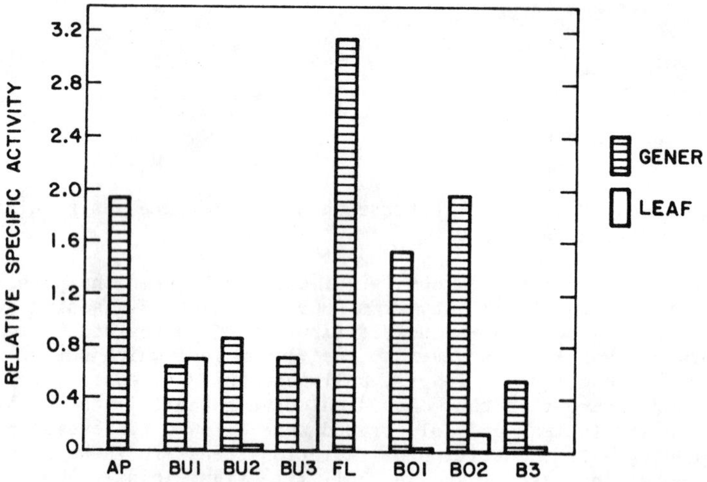

Figure 4 Translocation of ^{14}C from flower leaf to apex, buds, flower and bolls or nearby leaves in a fruiting branch of a cotton plant pruned to four fruit branches. Relative specific activity found 48 h after application. (6.9% R.S.A. was found in the flower leaf, the source leaf). AP = Apex, BU1-3 = bud 1-3, FL = flower leaf, BO1-3 = boll 1-3, GENER = generative organ

When the bud leaf was the source leaf, the relative specific activity increased at the upper part of the branch. Forty-eight hours after applying ^{14}C to a leaf, approximately 40% photosynthate produced was retained in the leaf. Buds or bolls kept nearly 90% of the photosynthate produced by themselves. On the other hand, the apex, although generally believed to be a strong sink, kept only 40-50% of its own photosynthesis. In all instances, the organ nearest to the source leaf produced the highest sink activity. These results of translocation in a fruit branch were produced in pruned plants, in which all leaves and generative organs received full sunshine radiation.

Placing one branch of a plant, pruned to two fruit branches, in plant shade or in artificial shade reduced the ability to produce buds or bolls considerably. As shown in Figure 5, when the 'inner' branch was in plant shade hardly any buds were produced, the apex degenerated but a few bolls were produced.

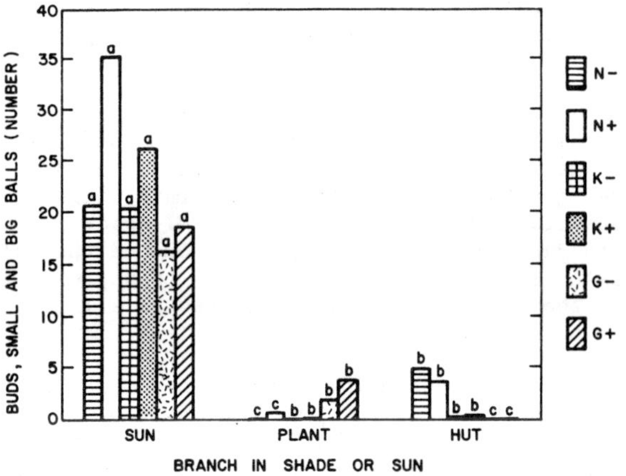

Figure 5 Development of buds and bolls on the 'inner' fruiting branch of a cotton plant pruned to two fruiting branches. The 'outer' branch was always in sunshine, the 'inner' branch was left in sunlight (SUN), in plant shade (PLANT), or in shade-net and paper - produced shade (HUT). N = bud, K = small boll, G = big boll, N+, K+, G+ are buds or bolls produced after excising all buds and bolls of the opposite 'outer' branch

The degeneration of the sympodial branch apex could be observed a few days after the apex came into plant shade. Branches placed in the shade produced by plastic net and paper, left the apex intact, produced some buds but practically all bolls abscissed. Branches in full sunlight produced abundance of buds or bolls. Removing all buds on the second 'outer' branch resulted in some of the photosynthate being translocated to the opposite branch, but most of it was used to increase dry weight of the 'outer branch'. The increase in dry weight of the shaded branch did not change the abscission rate and branches under paper shade did not produce bolls.

These experiments demonstrate that, despite the high transfer ability in the cotton plant, a nearby sink has priority over farther away ones. Furthermore, the sink activity and therefore the strength of sinks of buds, flowers and small bolls are inferior to those of large bolls. In addition, the influence of radiation on the phytochrome and plant growth substances which control the plant's structure and function should be noted (Smith, 1982).

To reduce bud and boll abscission and increase boll development, the generative organs must be as near as possible to highly active leaves. Radiation reaching the apex, buds or bolls themselves is not only beneficial but often also critical for the survival of the organ.

To increase cotton crop yield, the efficiency of the radiation must be fully utilised throughout the season by young and active leaves, but, on the other hand, radiation should reach the areas where buds and bolls develop. Therefore some compromise must be reached. An optimal growth rate may guarantee the necessary new active leaf area and yet allow ample radiation to reach the positions in the plant where buds and bolls develop.

Cotton (Acala SJ2) was grown at Mevo-Hama at 1 m intervals between rows with 8 ± 2 plants per m row. At the stage before boll load developed, the water potential of the plant was lowered to a level assuring planned growth rates. As the growth rate decreased, the water potential of the plant was raised to that level which produced the necessary growth rate. As the season progressed and the boll load increased, growth decreased in all treatments to approximately 1 cm per day. Therefore, the water potential was raised in all treatments to maximise photosynthesis and produce some new leaves and flower buds.

Although reducing water potential always decreased total dry matter production, the high harvest index produced maximum cotton yield and increased quality and water use efficiency. Figure 6 illustrate the influence of different growth rates on the daily-produced, total dry matter, cotton yield, earliness, boll weight, fiber ration and harvest index. The 25 mm-per-day growth treatment produced maximum dry matter but the lowest fiber yield and harvest index. The harvest 20 days after the first bolls opened produced 25 g cotton per m^2 per day for the 10 mm-per-day growth rate, while a growth rate of 25 mm-per-day produced a third of this amount, demonstrating the earliness of the latter treatment. The growth rate of 15 mm-per-day showed less earliness but gave a high total yield. The optimal growth rate has to be adapted to the specific growing conditions. An insufficient growth rate may result in an early 'cut-out' wasting part of the season and producing an inefficient regrowth at the end of the season.

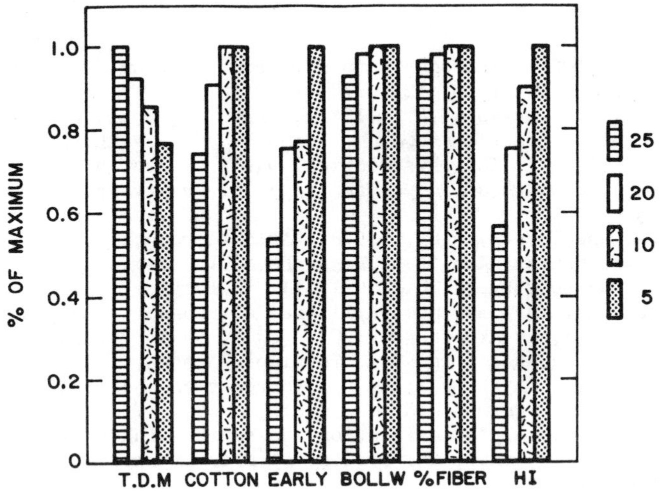

Figure 6 Cotton growth rate experiment, 1982. Treatments had daily growth rates between 5 and 25 mm due to changes of water potential of plant. Relative values for the different treatments. Maximum total dry matter produced = 1,958 g/m^2, cotton fiber yield = 308.2 g/m^2, cotton weight per boll = 7.40 g, fiber percentage = 39.3%. Figure shows % of maximum for the different treatments, 25, 20, 10 or 5 mm-per-day growth rate. TDM = total dry matter, COTTON = cotton yield, EARLY = earliness, BOLLW = boll weight, % FIBER = % of cotton fiber, HI = harvest index

SUMMARY

Water stress decreased total dry matter production while radiation stress reduced mainly the harvest index. To produce high yielding cotton crops with good fiber quality and high water use efficiency, an optimal radiation regime must be achieved in the cotton canopy. Radiation on and near flower buds and bolls reduced abscission and increased cotton weight and quality. These results can be explained by the influence of radiation on a number of physiological activities:

1. Increasing photosynthetic activity of the leaf.
2. Photosynthesis efficiency of the bud or boll.
3. Supply of nitrogen, minerals and growth substances from the root increases in sun-lit leaves with high stomatal conductance.
4. The position of a sink near leaves of high supply potential guarantees priority in the competition with other organs for photosynthate demand.
5. Sink activity increases with higher temperature. When positioned in direct sun radiation, bud temperature increased 5°C and boll temperature up to 8°C.
6. Many plant growth substances and important enzymes, in the photosynthesis or nitrate reduction cycles, increase activity or synthesis under high radiation fluxes.

REFERENCES

Bjorkman, O., (1981), 'Responses to different quantum flux densities', *Encyclop. Plant Physiol.*, 12A: 57 - 107.

Bradford, K.I. and Hsiao, T.C., (1982), 'Physiological responses to moderate water stress', *Encyclop. Plant Physiol.*, 12B: 263 - 324.

Evans, L.T., (1975), 'The physiological basis of crop yield', in Evans, L.T., (ed.), *Crop Physiology*, Cambridge University Press, Cambridge, 327 - 355.

Fisher, R.A. and Turner, N.C., (1978), 'Plant productivity in the arid and semiarid zones', *Ann. Rev. Plant Physiol.*, 29: 277 - 317.

Nobel, P.S. and Longstreth, D.J., (1981), 'Effects of environmental factors on leaf anatomy, mesophyll and cell conductance and photosynthesis', in Akoyunoglov, G., (ed.), *Photosynthesis*, Akoyunoglov Balaban Intern. Sci. Services, Athens 4: 245 - 254.

Smith, H., (1982), 'Light quality, photoreception and plant strategy', *Ann. Rev. Plant Physiol.*, 33: 481 - 518.

Wardlaw, I.F., (1980), 'Translocation and source-sink relationship', in Carlson, P.S., (ed.), *The Biology of Crop Production*, Academic Press, New York.

13 Contribution of agricultural engineering to optimal yield management of vegetables

YOAV SARIG

INTRODUCTION

In the report of the Club of Rome's project on the predicament of mankind, a gloomy picture is portrayed for future world economic and population growth (Meadows *et al.*, 1972). The aim of this report was to bring new ideas, new analyses and new approaches to persistent problems, to the attention of those who help determine the quality and direction of our life. It calls for people in all fields of endeavor to consider the need for concerted action now, if we are to preserve the habitability of this planet for ourselves and our children.

One of the major factors in the category of physical necessities that are required to sustain world economic and population growth is the production of food. The authors of the report state the limits to food production, but also maintain that it is possible to avoid or extend these limits by technological advances that remove dependence on natural limitations.

Optimal yield management addresses itself to this commitment of pushing back the apparent limits to food production. Nevertheless, overcoming these limits cannot be perceived without the contribution of modern technology.

Agricultural mechanisation or agricultural engineering in the broader sense, albeit not the predominant technological factor, plays an important role in achieving optimal yield objectives. Quite often, and mistakenly, agricultural mechanisation is overlooked with respect to more traditional factors, such as water, land, seeds and fertilisers. This approach is justified if optimal yield is interpreted only as maximal yield, since these factors are commonly associated with increasing the quantity produced. However, in modern agriculture optimal yield should be envisaged in a much broader sense, taking into consideration the abundant factors of production. In this multifaceted approach the contribution of agricultural mechanisation is significant, both directly and indirectly, as will be demonstrated for the production of vegetables, as an example.

DIRECT CONTRIBUTION

The assertion that mechanisation leads directly to increased yields has been shown to be true only when associated with other production factors, such as high-yielding seeds, pesticides and fertilisers (Binswanger, 1986). Thus, the introduction of further mechanisation is usually a substitute for labour.

This generalisation which represents the 'substitution view' of agricultural mechanisation (Binswanger, 1986) differs from the net contribution view, which assumes that more machinery produces higher yields or other gains in output, regardless of the economic environments in which it is introduced. For example, yields may be low in an extensive vegetable growing area in Africa, where hoe cultivation is used, while in an intensively farmed tractorised region of India yields may be much higher (Binswanger, 1986). The yield differences, however, may be caused, in part, by differences in other inputs, such as fertilisers or seeds. They could also be caused by better tillage in India - but this does not mean that good tillage is achieved only by tractors and can not be achieved by hand. Examples from Java and from the Arab farm sector in Israel have shown that cultivation by hand can be as thorough as by tractor (Binswanger, 1986). The lower tillage intensity in Africa may simply reflect the abundance of land: in order to maximise labour productivity, people work sparsely over a large area.

From the example of Africa we can derive the next generalisation namely that mechanisation is most profitable and contributes most to growth where land is abundant, and/or labour is scarce relative to land, and/or where labour is moving rapidly off the land. The validity of this generalisation can be illustrated by tomatoes grown for processing in the USA. The USA is a worldwide leading producer of vegetables for processing, in general, and tomatoes, in particular (Portas, 1983). This is primarily the result of the introduction of mechanisation motivated by labour scarcity on the one hand (due to restrictions on migratory labour), and availability of land on the other.

It should be noted, and can be clearly seen in Table 1, that although mechanisation has led to massive output growth, higher yields came only later and were linked to fertilisers and biological innovations.

Table 1

Direct and indirect effects of agricultural mechanisation

Factors leading to mechanisation	*Immediate consequence of mechanisation*	*Indirect effect on agricultural putout*	*Indirect effect on agricultural employment*	*Examples*
1. Land availability	Labour used on larger areas, production costs drop	Expands, and more quickly the more elastic is final demand	Expands if demand is elastic; stagnates or falls if demand is inelastic	Nineteenth century United States
2. Wages rising in response to nonagricultural labour demand	Production costs rise less than in absence of mechanisation	Falls (or grows more slowly), but by less than in absence of mechanisation	Falls	United States after 1940, Japan, Europe after 1955
3. Nonmechanised technique unprofitable	A new method of production becomes profitable	Expands, and more quickly the more elastic is final demand	Expands, and more quickly the more elastic is final demand	Pumping in contemporary Asia
4. Subsidies on capital energy	Production costs may drop modesty or stay constant	Small expansion at best	Falls, sometimes sharply	Contemporary Brazil, Pakistan, China

Source: Binswanger, 1986.

Another illustrative case is the production of paprika in Israel, depicting Case 3 in Table 1. Although it is always possible to pick paprika by hand, it is not feasible on a large acreage, even at low wages, because of low output resulting from the need for selective harvesting. It is definitely unprofitable in the case of high wages, resulting from nonagricultural labour demand. The introduction of mechanical harvesting of paprika was a powerful stimulus to the growth of this crop and paved the way for establishing a viable processing industry for paprika in Israel (Wolf and Alper, 1983).

Rhubarb production provides an example depicting Case 2 in Table 1 illustrating the contribution of mechanisation. Thus production of rhubarb in the USA has declined and the number of growers reduced because of high relative harvest cost based on conventional hand labour. The development of a mechanical harvester has resulted in savings of up to 60 percent in harvesting cost even though recovery decreased (Marshall, 1983).

In order to maximise the yield of peanut production it is necessary to recover the peanuts which remain buried in the ground after the conventional harvesting, both manual and mechanical. Under certain conditions this portion can be as high as 30 percent! A peanut gleaner was developed in Israel for both light and heavy soils, resulting in 98 percent recovery, thus maximising the net profit for the grower (Mizrach *et al.*, 1983). Direct contribution of agricultural engineering in the broad sense can further be illustrated in numerous examples.

Growing tomatoes in greenhouses can produce fruit of high yield and quality. However, it requires maintaining a proper environment for plant growth - a very intricate process incorporating diversified engineering input (Ben-Jaacov *et al.*, 1980). Another direct contribution to the production of greenhouse tomatoes is the mechanisation of pollination, which constitutes the largest labour-consuming activity. Although tomatoes are considered self-pollinating, if left to themselves only about 30% of the bloom normally set fruit. Greenhouse growers have found that by mechanically agitating the bloom, a satisfactory set can be obtained (DeTar *et al.*, 1967). A technique based on a pulsating air jet incorporated in machines for both greenhouse tomatoes and open fields resulted in a higher mean weight per tomato and significantly increased yield (Nahir *et al.*, 1986), mechanisation does not necessarily imply complex and sophisticated machines.

Small farms that depend on family labour can obtain high profit from intensive farming of selected vegetable crops. However, these producers, predominate in developed countries, are experiencing increasing difficulty in finding harvest labour. The drudgery of hand harvesting of vegetables has caused a decline in available hired labour and has served as a deterrent to the adoption of certain crops by small family farms. This unfortunate trend is manifested, for example, in harvesting cucumbers in Kentucky (Walton *et al.*, 1986), vegetable growing in Israel (Alper and Wolf, 1983), cabbage harvesting in Florida and also in vegetable growing in the USA and many countries in Europe (Bennedsen, 1983; Holmes and Halsen, 1959; Lenker, 1983). In all these countries low-cost machines are needed to facilitate production and to make harvesting easier and faster. It also enables an optimal timing of harvesting - a crucial factor for certain crops such as cucumbers and tomatoes, which ripen rapidly and require multiharvesting for optimal marketability.

Various mechanical aids have been developed in most intensive vegetable growing regions, generally resulting in higher harvesting rates than by conventional hand harvest. The major advantages of most picking aids lie in their continuous movement (and therefore continuous picking), the elimination of bending for harvest, effective use of both hands for harvest and carrying the produce on the machine - all of these features reduce fatigue, thus optimising the output of the workers.

Optimisation in the above examples is closely related to other factors which indirectly contribute significantly, to the overall optimisation of the process.

The relatively low cost of simple devices lends their use to small farmers and thus increases profitability. Moreover, the simplicity of design enable then to be used and maintenanced by less skilled workers.

INDIRECT CONTRIBUTION

Optimisation of yield in its broader sense may also be construed as quality optimisation of the produce.

Achieving maximum yield of tomatoes as a result of high yielding seeds, use of fertilisers and irrigation systems would not necessarily maximise profit if the quality of the tomato is low. Low-grade tomato can be the result of poor sorting, inaccurate sizing, uneven firmness and mechanical damage.

Mechanical sorting, for example, based on the difference in the physical properties of red and green tomatoes and difference in firmness is less biased, more consistent and less labour involved than hand sorting. Hence, higher grade tomatoes are marketed and better utilisation of the product obtain by chanelling the various grades to optimal and use (e.g., less firm tomatoes will be processed as juice or paste, and the firmer ones as peeled).

Hence, a knowledge is required of the basic physical properties of the tomato to provide the essential engineering data for the design of machines, structures, processes and controls and for evaluating and retaining the quality of the final product (Mohsenin, 1970). Maintenance of quality under adverse conditions of handling, storage and distribution, saving in weight and bulk in some cases, reduced costs in handling and processing operations, and finding new ways for utilisation may result from an understanding of the physical properties of economic plants. It should be noted that such basic information is of value not only to engineers, but also to food scientists and processors, plant breeders and other scientists who may exploit these properties and find new uses.

During the past twenty years, the subject of physical properties of agricultural products and their contribution to the improvement of many agricultural processes has advanced significantly. Many fields of interest have been identified and include determining the physical state of materials and elucidating mechanical, thermal and electromagnetic properties, and potential practical implementation of the knowledge obtained. Numerous examples can be shown in almost all facets of agriculture (Mohsenin, 1970).

Another indirect contribution of mechanisation in modern agriculture is the application of pesticides, which constitute an essential component of crop protection. The use of hand-held and hand-operated pesticide applicators is impractical over large areas. In addition, the amount of plant protection chemicals used in most developed countries is increasing continuously, leading not only to enhanced production costs, but also to increased exposure to hazardous materials. Furthermore, inadequate pesticide coverage of the plant (primarily lower leaf surfaces) render insufficient insect and disease control which leads to fruit infection resulting in commercial losses.

The quantity of a pesticide required per unit area and its control efficacy are greatly affected by the application machinery, technique and process. Developments and adaptation of new application techniques and equipment would permit savings in costly chemicals and minimise environmental pollution.

FUTURE TRENDS

Optimal yield management in its broader sense encompasses many facets in agriculture and is the result of integrated inputs. Thus, although a wide gap exists at present between developed and less developed countries, one has to take into consideration future inputs of new technologies and their contribution and impact on optimal yield management.

One of the newer methods that utilises high technology and that contributes both to technical and tactical management is the use of expert systems. The idea behind the method is that the computer system is programmed to simulate an expert in communication with a client who has a problem to be solved. Thus, the knowledge of expert individuals is harnessed in computer routines so that relatively unskilled persons can run sophisticated programs. Expert systems are already in use in several industries, and their use in agriculture is imminent. Applications which have already been proposed, and partially even implemented, relate to queries on diagnosis, control problems and tactical,short-term management problems (Peart *et al.*, 1986).

The introduction and use of new technologies (known in the trade as 'high tech') is not limited to developed countries. Although developing a new sophisticated system, incorporating new scientific breakthrough, may entail a high level of technology, their use may be simple enough to attract a wide range of clientele. (An expert driver does not necessarily understand how the motor of his car operates).

Another area that is likely to be a major area of research is the development of dedicated sensors. Human sensing, albeit an impressive technology by itself, has distinct shortcomings, thus limiting the optimisation of many processes. Nondestructive techniques to ascertain internal quality attributes of many agricultural products are impossible with human sensing only. Determination of sugar content in melons, internal rot in peaches, hollowness in cucumbers or black heart in potatoes are only a few examples where sensors based on recent breakthrough in high tech can provide the necessary knowledge for improved quality and hence optimal yield.

Again, the obvious trend of relying on objective, unbiased and consistent mechanical sensors capable of operating under hostile environment and generate reproducible results is not limited to the developed countries. It represent the philosophy of modern agriculture which requires an ever increasing use of advanced technology to fulfill its major goal of food production.

REFERENCES

Alper, H. and Wolf, I., (1983), 'Mechanical harvesting of vegetable in Israel', *Proc. 1st Int. Symp. on Fruit, Nut and Vegetable Harvesting Mechanization*, ASAE Publication No. 5-84: 34 - 39.

Ben-Jaacov, J., Hagiladi, A., Levav, N. and Zamir, N., (1980), 'The Hydrosolaric greenhouse - a new growing and propagating environment', *Proc. Int. Plant Propagation Society*, 30: 366 - 371.

Bennedsen, B.S., (1983), 'Mechanical harvesting of vegetables in Denmark', *Proc. 1st Int. Symp. on Fruit, Nut and Vegetable Harvesting Mechanization*, ASAE Publication 5-84: 22 - 24.

Binswanger, H., (1986), 'Agricultural Mechanization - A comparative historical perspective', *Research Observer*, 1: 27 - 55.

DeTar, W.R., Haugh, C.G. and Hamilton, J.F., (1967), 'Accoustically forced vibration of greenhouse tomato blossoms to induce pollination', ASAE Paper No. 67-817, Presented at the ASAE Winter Meeting, Detroit, MI.

Holmes, E.S. and Halsen, L.H., (1959), 'Harvesting cabbage with mechanical aids in Florida, *Fl. Agr. Exp. Stat. Circular S.*, 114: 14.

Lenker, D.H., (1983), 'Factors limiting the harvest mechanization of some major vegetable crops in the USA', *Proc. 1st Int. Symp. on Fruit, Nut and Vegetable Harvesting Mechanization*, ASAE Publication No. 5-84: 29 - 34.

Marshall, D.E., (1983), 'Harvest mechanization of field-grown rhubarb', *Proc. 1st. Int. Symp. on Fruit, Nut and Vegetable Harvesting Mechanization*, ASAE Publication 5-84: 260 - 265.

Meadows, D.H., Meadows, D.L., Randers, J. and Behrews III, W.W., (1972), *The Limits to Growth*, Potomac Associates, Wash. D.C.

Mizrach, A., Margolin, A., Feller, R. and Alper, Y., (1983), 'Peanut salvage machine for sandy loam and clay soils', *Trans. ASAE*, 26: 389 - 391.

Mohsenin, N.N., (1970), *Physical Properties of Plant and Animal Materials*, Gordon and Breach Science Publ, N.Y.

Nahir, D., Ganmor, S., Rylski, I. and Frankel, H., (1983), 'Pollination of tomato flowers by a pulsating air jet', *Trans. ASAE*, 27: 896 - 899.

Peart, K.M., Zazueta, F.S., Jones, P., Jones, J.W. and Mishoe, J.W., (1986), 'Expert systems take on three tough agricultural tasks', *Agricultural Engineering*, 67(3): 8 - 10.
Portas, C.M., (1983), 'Mechanical harvesting of vegetables for processing: A world-wide view', *Proc. 1st Int. Symp. on Fruit, Nut and Vegetable Harvesting Mechanization*, ASAE Publication 5-84: 39 - 45.
Walton, L.R., Swetman, L.D. and Casada, J.H., (1986), 'A harvesting aid for picking cucumbers', *Applied Agriculture*, 2(11): 41 - 44.
Wolf, I. and Alper, Y., (1983), 'Mechanization of paprika harvests', *Proc. 1st Int. Symp. on Fruit, Nut and Vegetable Harvesting Mechanization*, ASAE Publication 5-84: 265 - 276.

14 COTMOD – a model to optimise cotton response to irrigation and nitrogen

AVISHALOM MARANI

INTRODUCTION

The use of simulation models of crop systems is a new and sophisticated tool for more efficient management of crops. The subject has been reviewed recently by Whisler *et al.* (1986). Several models of the cotton crop have recently been developed, the most comprehensive among them being GOSSYM (Baker *et al.*, 1983). An attempt has been made to adapt this model for the Acala type cotton, grown under irrigation (Marani and Baker, 1981).

Although the performance of the GOSSYM model was found to be acceptable for intensively grown cotton under irrigated conditions, it became apparent that a new model, which will be more sensitive to moisture stress and to nitrogen deficiency, is required. In response to this need, the COTMOD model has been developed.

DESCRIPTION OF THE MODEL

COTMOD is a process-level model of the cotton crop, based on available scientific information on physical and physiological processes in the soil and plant, and on the microclimatic environment. It is a materials-balance model, and it deals quantitatively with the carbon, water and nitrogen in the soil - plant - atmosphere systems. The soil is divided into horizontal layers, and plant parts are arranged in arrays according to the dates of their formation.

The program is written in FORTRAN 77, and it can be run on any mainframe or minicomputer with a standard FORTRAN 77 compiler, as well as on IBM PC compatible microcomputers fitted with an 8087 mathematical coprocessor. The model requires as input daily climatic data consisting of minimum and maximum temperatures, global radiation, rainfall and class A pan evaporation. Actual meteorological measurements are used for past periods, and perennial average data are used for future predictions. The model also requires as input soil data consisting of field capacity, permanent wilting point, initial moisture content, initial nitrate and ammonium nitrogen contents, soil organic matter, clay content and bulk density. The user is also required to input data of the cultural methods used in the particular field: dates of planting and germination, dates and amounts of irrigations and nitrogen applications, row spacing and plant density within the row.

The model consists of four main parts:

1. The main program and subroutines for managing input and output files, reruns of the model when necessary, and to initial the state and rate variables. Most of the processes are calculated on a daily time-step basis, and each day the following three sections are called.
2. The climate section. Climatic data are read or estimated for each day, and the necessary parameters needed for the phenological and physiological functions are calculated.
3. The soil section simulates the temperature regime in the soil and within the canopy, water and nitrate movement between soil layers, evaporation of water from the soil surface, rate of transpiration, mineralisation of soil organic nitrogen, nitrification and denitrification processes and uptake of nitrogen by the plant. Some of the soil processes are calculated on the basis of shorter time-steps (of 1 to 6 h).
4. The plant section simulates the phenology of the cotton plant, photosynthesis and respiration, plant height and leaf area, the formation, growth and abscission of fruiting parts, flowering and boll opening. Dry matter or carbohydrate balance is the result of 'source' (= photosynthesis - respiration) and 'sink' (= requirements for potential growth) relationships, and it affects the possible occurrence of carbohydrate stress. Nitrogen in the plant parts is simulated as the result of uptake from the soil and the requirement for the actual growth of each part. When supply is less than the demand, nitrogen stress is developed.

EVALUATION OF MODEL PERFORMANCE

The model has been tested by using many sets of field data, and comparing the predictions of the model with actual results. This process is called 'validation'.

As an example, let us take an experiment conducted at Hatzor (in the coastal plain of Israel) in 1982. The climate at this location is rather mild (average maximum and minimum temperature being 30° and 18°C, respectively, during the middle of the summer) and without any rainfall from April to October. The soil is a deep alluvial clay loam with good water holding capacity. There were 18 different irrigation treatments in this experiment, consisting of one to six irrigations applied at different periods and with different amounts of water. We tried to simulate the effects of these irrigation treatments on the lint yield, and plotted the predicted values against the real values (Figure 1).

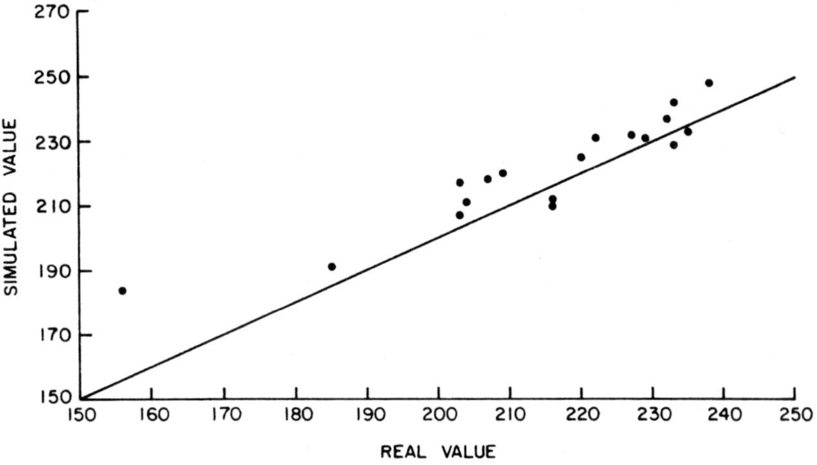

Figure 1 Hazor Irrigation Exp., 1982, results of simulation by COTMOD

Most of the points in the figure were very near to the 1:1 line, indicating that the model simulated very accurately the responses to the different irrigation regimes.

It is very important to examine the usefulness of any model under a variety of climatic and soil conditions. We have therefore tried to test the model by using it for environments that differ from those in Israel. We had available data from two cotton regions in South Africa. In one region (Upington, northern Cape Province), which is a desert area with low rainfall and a clay loam soil, the need for three irrigations was predicted. In the second region (Loskop, central Transvaal), an area with high rainfall and a sandy loam soil, it was predicted that only one irrigation was needed during the season.

This indicates that the model can also simulate cotton production in rain-fed regions. Rainfall is generally treated by the model in a similar manner to irrigation, but problems such as the effect of rain, runoff or drainage should be taken into consideration.

ON-FARM USE OF THE MODEL

The model is not complete, and there still remains much work to develop many of its aspects. However, the present incomplete version is useful to the cotton manager for analysis of some complicated situations. For example, when we applied the model to the Loskop data, we found that the predicted yields were very low (and in fact they are actually low in this area). Analysis of the results of the model indicates two main reasons for this: (1) low radiation levels during many cloudy days; and (2) nitrogen stress developed during the late part of the season, due to leaching and denitrification. Running the model with additional late-season applications of nitrogen improved the yield to a certain degree, indicating an effective practice that can be adopted by the growers under these conditions.

The model may be used for crop management by predicting possible crop development under various environmental or management conditions. For irrigation management, it is very important that accurate values for soil constants and for soil moisture at the beginning of the simulation will be input by the user. The soil moisture content of each soil layer is calculated each day by simulating drainage of rain or irrigation water, evapotranspiration, and water transport between soil layers by diffusion. Soil moisture in relation to plant root activity, and climatic conditions, determines the degree of moisture stress in the plant. The model simulates the effect of moisture stress on plant growth growth, photosynthesis, formation of fruiting parts, allocation of photosynthates to the bolls, abscission of fruiting parts and leaf senescence. Thus, the effect of irrigation regimes on final lint yield, under different environmental conditions, may be predicted by the model.

Several tentative irrigation regimes may be defined by the user and evaluated before planting, by running the model for a single set or several sets of environmental conditions. The model will predict the probable yield under several possible weather conditions. Thus, some of the irrigation plans will be rejected, and others will be recommended for use. During the growing season, the model may be used to reevaluate and adjust the actual irrigation applications.

Alternatively, the model may be set to predict automatically the optimal irrigation regime under the given soil and climatic conditions, using scientifically established and commonly accepted criteria.

The problem of nitrogen management is approached similarly. The user inputs the initial soil N content, as determined from soil analyses, as well as a suggested sequence of fertiliser applications. Processes of nitrogen mineralisation and nitrification, transport of nitrates between soil layers, uptake of nitrogen by the plant, as well as the effects of nitrogen stress on cotton growth and fruiting, are simulated by the model. Thus, the expected yield of lint under these conditions is predicted.

Different fertilisation regimes may be defined by the user, and their effect on cotton yield will thus be evaluated for given soil, climate and cultural treatments, by using the model. Taking into account the predicted effects on lint yield and quality, as well as on crop earliness, the optimal nitrogen fertilisation may be determined.

It should be emphasised that the efficient use of any model depends on the availability and accuracy of climatic and soil data. Soil data, such as water holding capacity parameters or available nitrogen content at the beginning of the season, may be obtained by sending soil samples to suitable laboratories. Climatic data may be obtained from regional weather stations, or from on-site weather stations in large farms. At the beginning of the season, when the climate is unknown, perennial average data or several possible previous 'climate scenarios' may be used for simulation. As the season progresses, more accurate predictions may be made by using actual weather data up to a certain date.

REFERENCES

Baker, D.N., Lambert, J.R. and McKinion, J.M., (1983), 'GOSSYM: a simulator of cotton crop growth and yield', *S.C. Agric. Exp. Stn. Tech. Bull.*, 1089: 134.

Marani, A. and Baker, D.N., (1981), *Development of a Predictive Dynamic Simulation Model of Growth and Yield in Acala Cotton*, U.S. - Israel Binatl. Sci. Found. Tech. Res. Rep., 177.

Whisler, F.D., Acock, B., Baker, D.N., Fye, R.E., Hodges, H.F., Lambert, J.R., Lemmon, H.E., McKinion, J.M. and Reddy, V.R., (1986), 'Crop simulation models in agronomic systems', *Advances in Agronomy*, 40: 141 - 208.

15 Double-cropping system (cotton–wheat) based on frequent drip irrigatin and control of the root zone system

AVNER CARMI AND ZVI PLAUT

INTRODUCTION

Cotton is a crop for which the possibility of growth control has a major agronomic importance. Excessive vegetative growth may be followed by yield reduction (Jackson and Tilt, 1968; Grimes *et al.*, 1969). The interrelationships between vegetative and reproductive growth results from competition for assimilates between both processes (Eaton and Ergle, 1953). Irrigation management of cotton strongly affects shoot growth rate (Bunce, 1978; Bielorai *et al.*, 1983) and the distribution of assimilates between its vegetative and reproductive organs (Grimes *et al.*, 1982). Careful control of root zone humidity is a basic means by which the grower can affect growth processes. Efficient control of root zone humidity can be achieved by the use of a drip irrigation system due to its accurate application of water, ease and flexibility in altering irrigation amounts and intervals and the vicinity of the drippers to the irrigated root zones (Bresler, 1977; Dasberg and Bresler, 1985). Carmi and Shalhevet (1983) reported that restriction of cotton root zone volume, in pots, under management of frequent pulses of irrigation restrained vegetative growth, accelerated reproductive development and shortened the growth period. Carmi and Plaut (1982) applied that concept using a trickle irrigation system to affect the establishment of a restricted root zone under practical field conditions. The specific irrigation management used enabled the sizes of the wetted root zone and its water content to be altered easily, and thereby controlled growth processes and affected earliness. The possibility of achieving high cotton yields in a short growth period opened the way to a double-cropping system, namely cotton -

wheat. In such a system, a short growth period of the two crops is a precondition needed to prevent an overlapping in their growth periods.

The purpose of the present work was to study the possibility of using drip irrigation management as a tool to establish a restricted root zone in cotton and to evaluate its agronomic performance.

MATERIALS AND METHODS

Cotton seeds *(Gossypium hirsutum L.* 'Acala SJ-2') were planted in Nir-am, at the end of May following wheat which was harvested in mid-May. The soil (loess type) was predried by the winter crop. An irrigation of 30 mm was given during the process of soil tillage and another 30 mm applied by sprinklers for germination. Plant stand was 100,000 per hectare grown in row spacing of 100 cm. The trickle irrigation network consisted of drippers (discharge 1.4 L/h) in lines 40 cm apart. The dripper lines were located between two adjacent plant rows, at distances of 200 cm apart. Trickle irrigation started about 14 days after seedling emergence. Three levels of irrigation were used. Distribution of water supply during the growth period and its amount is given in Table 1.

Table 1

Average daily evaporation rates of class A pan and daily amounts of water supplied by drip irrigation. Plants were irrigated with the following overall* amounts of water: 460 mm (low), 511 mm (intermediate) and 550 mm (high). Plants were sown on 6 June

Period	*Average daily pan A evaporation rate (mm)*	*Average daily irrigation rate (mm)*		
		Low	*Intermediate*	*High*
June 23 - 30	6.4	2.6	2.6	2.6
July 1 - 5	6.4	2.9	2.9	2.9
6 - 12	6.6	3.8	3.8	3.8
13 - 19	7.0	4.0	4.7	5.2
20 - 26	6.5	4.2	4.9	5.9
27 - 30	6.8	4.2	4.9	5.9
Aug. 1 - 3	6.9	4.2	4.9	5.9
4 - 10	6.1	4.8	5.5	6.2
11 - 17	6.7	5.2	5.7	6.3
18 - 24	6.1	5.7	6.6	7.3
25 - 31	6.2	5.7	6.6	7.3
Sept. 1 - 6	5.0	5.5	6.0	6.5
7 - 13	5.4	5.0	5.6	5.6
14 - 21	4.6	3.4	4.0	4.0
22 - 28	4.4	2.0	2.0	2.0

* Overall amount included tillage wetting and germination irrigation.

Evaporation measurements from a class A pan were taken as an indicator of the meteorological effects on transpiration rate and are presented in Table 1. Soil water content in the 0 to 80 cm soil layer was measured weekly during the growth period by a neutron probe. The measurement sites were located 20 cm away from the dripper towards the plant row. Nutrients were consistently supplied with the water through the irrigation system, in quantities detailed in Table 2.

Table 2

Monthly amounts of nitrogen and phosphorus supplied to treatments irrigated with the following overall amounts of water: 460 mm (low), 511 mm (intermediate) and 550 mm (high). Plants were sown on 6 June

	Fertilisation rate (kg/ha)					
Period	*Nitrogen*			*Phosphorus*		
	Low	*Intermediate*	*High*	*Low*	*Intermediate*	*High*
June	130.0*	130.0*	130.0*	7.6	7.6	7.6
July	86.4	89.0	106.8	21.6	22.2	26.7
Aug.	105.4	114.3	129.1	26.3	28.5	32.2
Sept.	63.0	71.9	71.9	15.7	17.9	17.9
Total	384.8	405.2	437.8	71.2	76.2	84.4

* 100 kg/ha of that amount was supplied at time of soil tillage to correct for the deficiency in soil content caused by the wheat.

Each treatment consisted of four replicated plots of 6 m width and 8 m length. Cotton was picked manually.

RESULTS

Measurements of available water in the soil profile are presented in Table 3.

Table 3

Content of available water in the soil as calculated from neutron probe measurements. Available water was calculated on the basis of range humidity between wilting point and field capacity. Plants were irrigated with the following overall amounts of water: 460 mm (low), 510 mm (intermediate) and 550 mm (high)

	Soil depth (cm)											
	0 - 30				*30 - 60*				*50 - 80*			
Irrig. rate / *date*	*21/7*	*8/8*	*31/8*	*21/9*	*21/7*	*8/8*	*31/8*	*21/9*	*21/8*	*8/8*	*31/8*	*21/9*
Low	27	44	61	72	21	13	29	56	4	0	4	20
Intermediate	58	35	77	80	32	6	44	64	21	0	29	32
High	44	61	44	65	38	26	36	44	26	15	26	20

Most of the available water supplied by irrigation were found in a shallow subsurface layer of 0 to 30 cm depth. A smaller amount was available in the 30 to 60-cm layer and almost none below depth 50 cm. During the last month of irrigation, after 31 August, increased infiltration of water to deeper layers did occur. At that time a considerable decrease in evaporation rate was detected (Table 1), which was not accompanied by a proportional decrease in irrigation rate, resulting in a deeper wetting of the soil profile. The combined effect of frequent irrigations with small amounts of water was responsible for the high level of available water in the 0 to 30-cm depth. The water application rate ranged around 50% of maximal available water content of the specific soil type. Increasing irrigation rates from 460 to 510 or 550 mm water increased the available water in all the soil profiles examined and the infiltration of water to deeper layers (Table 3). Soil excavations were carried out with the aim of studying the effects of the specific irrigation management and the shallow wetting of the soil on root development and location. A photograph of a typical root system developed under the specific irrigation management is shown in Figure 1.

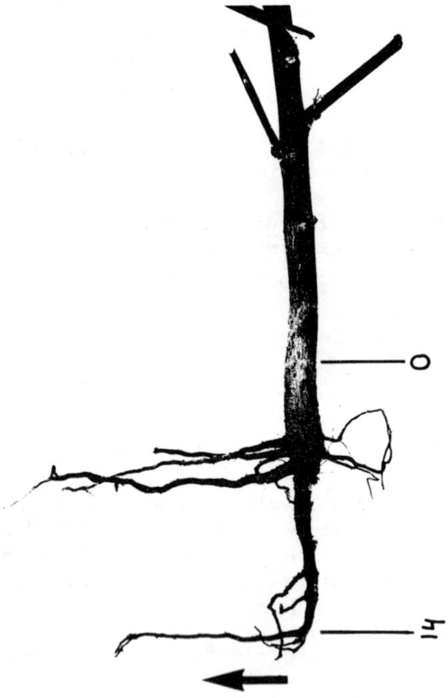

Figure 1 Photograph of plant root system 120 days after sowing. Arrow = direction of drippers line

It can be seen that root development took place towards the closest dripper line. The root system was shallow and most of the roots appeared in the 0 to 20-cm depth. A very high concentration of thin roots was observed in the vicinity of the drippers. Measurements of leaf water potential during the growth period are presented in Table 4.

Table 4

Leaf water potential (-MPa) of plants irrigated with the following overall amounts of water: 460 mm (low), 510 mm (intermediate) and 550 mm (high). Measurements were taken at midday

	Leaf water potential (-MPa)			
Irrig. rate *Date*	*21 July*	*28 July*	*17 August*	*6 September*
Low	1.87±0.08	1.65±0.03	2.34±0.02	1.68±0.05
Intermediate	1.50±0.02	1.21±0.02	1.37±0.05	1.26±0.02
High	1.28±0.03	1.36±0.02	1.56±0.04	1.27±0.06

Irrigation rates considerably affected leaf water potential: plants irrigated with the smallest quantities of water exhibited potentials ranging from -2.3 to -1.6 MPa compared to -1.3 to -1.5 MPa for plants that were highly irrigated. The dependence of plants on a daily supply of water was shown by the strong effect of a short delay in irrigation on leaf water potential. A 24-h delay in daily irrigation caused a very considerable decrease in leaf water potential, from -1.4 to -2.4 MPa, and this level remained stable for a further 24 h of an irrigation delay (Table 5).

Table 5

The effects of 24 or 48 h delay of a single daily irrigation on leaf water potential of 80-day-old plants. Measurements were taken at midday. IC = irrigation continued; IP = irrigation postponed

*Irrigation rate**		*Leaf water potential (-MPa) after an irrigation delay of*	
		24 h	*48 h*
Low	IC	1.41±0.04	1.40±0.05
Intermediate	IC	1.36±0.03	1.29±0.05
High	IC	1.37±0.05	1.32±0.04
Low	IP	2.42±0.19	2.23±0.04
Intermediate	IP	1.67±0.08	1.84±0.05
High	IP	1.67±0.13	1.80±0.04

* As in Table 3.

The effect was much less pronounced in plants which were under high irrigation management. Vegetative growth was strongly affected by irrigation rate. Increasing irrigation level enhanced shoot growth characterised by longer stems (Table 6) and higher shoot dry materials content (Table 7).

Table 6

Effect of irrigation rate on main stem height and internode number. Plants were irrigated with the following overall amounts of water: 460 mm (low), 510 mm (intermediate) and 550 mm (high)

Date	*6 July*		*14 August*		*14 September*	
Irrig. rate	*Stem height (cm)*	*Internode number*	*Stem height (cm)*	*Internode number*	*Stem height (cm)*	*Internode number*
Low	22.7±0.5	6.8±0.2	94.1±2.8	14.5±0.4	105.0±4.5	17.5±1.0
Intermediate	-	-	100.5±1.8	14.5±0.6	130.3±3.8	19.7±0.9
High	-	-	124.9±5.8	17.0±0.6	158.7±2.8	20.7±1.5

* Until 6 July, all treatments received the same quantity of water.

Table 7

Effects of irrigation rate on dry matter accumulation in shoot organs of 95-day-old plants. Plants were irrigated with the following overall amounts of water: 460 mm (low), 511 mm (intermediate) and 550 mm (high)

	Dry matter weight per plant (g)		
Irrig. rate	*Stem plus petioles*	*blades*	*total*
Low	40.6±2.1	35.4±1.6	76.0±3.5
Intermediate	58.1±2.6	44.5±3.5	102.6±3.5
High	68.2±4.9	50.1±1.9	118.3±6.6

The enhanced growth of the stem originated mainly from stem elongation while internode number was less affected (Table 6). The reproductive development was strongly affected by irrigation rate. An irrigation of 460 mm gave a seed cotton yield of 6,230 kg/ha which was picked significantly earlier compared to that which was irrigated with 510 mm (Table 8) and gave the later yield of 4,920 kg/ha.

Table 8

Effects of irrigation rate on yield and maturation. Plants were irrigated with the following overall amounts of water: 460 mm (10w), 510 mm (intermediate) and 550 mm (high)

Irrig. rate	*Seed cotton yield (kg/ha)*	*Cumulative yield (in percent of total yield)*			
		19 Oct.	*26 Oct.*	*2 Nov.*	*8 Nov.*
Low	6230±180	56±3	76±2	91±3	98±1
Intermediate	5560±320	46±2	68±3	82±2	92±2
High	4920±280	39±3	62±3	76±2	92±2

It seems that a progressive decrease in yield and delay of boll maturation and opening results from a progressive increase in irrigation level. A considerable decrease in flowering rate from around 183 opened flowers per square meter to a rate of 150 resulted from an increase in irrigation level from 460 to 550 mm (Table 9).

Table 9

Effects of irrigation rate on flowering and boll abscission rate. Plants were irrigated with the following overall amounts of water: 460 mm (low), 510 mm (intermediate) and 550 mm (high)

Irrig. rate	*Flowers/m²*	*Boll abscission (percent)*
Low	183±11	51±4
Intermediate	143±13	37±5
High	150±10	59±4

A large increase in boll abscission from 51 to 59% was an additional result of excessive irrigation (Table 9).

DISCUSSION

A major objective in the present work was to study the possibility of using a drip irrigation system as a tool which, in combination with specific irrigation management, will promote the establishment of a restricted root zone. Such a system, according to former research (Carmi and Plaut, 1982; Carmi and Shalhevet, 1983; Carmi, 1986), has the potential of better controlling the growth processes and increasing productivity. Klepper *et al.* (1973) found that keeping the upper soil profile well watered encouraged the development of the roots in the upper zones and restricted the penetration and growth of roots below the 100-cm depth. This was also noted in the present work. During most of the growth period, water was available to plants mainly in the 0 to 40 cm-depth (Table 3). Such conditions promoted root growth in that shallow wetted soil layer, as shown in Figure 1. Plant irrigation in contrast to the conventional irrigation concept (Grimes *et al.*, 1978) started soon after germination. The daily irrigation of small water pulses were supposed to build up a humidity gradient in which water content increased from the plant row zone towards the drippers. Under such conditions, root growth was promoted in the direction of the drippers (Figure 1) and a massive development of thin roots was observed in the humid zone around the drippers. A similar effect was observed by Willoughly and Cockroft (1974) who reported new root proliferation in the vicinity of drippers. The establishment of a shallow wetted root zone in the vicinity of the drippers enabled the root zone humidity not to be changed and consequently the plant water status had minor changes in the daily amount of water supplies. The difference in daily irrigation rate between low-water and high-water treatments did not exceed 0.6 to 1.7 mm (Table 1) during most of the growth period. However, such small changes were enough to cause considerable growth responses, as shown in Table 6 and 7. This indicates high water use efficiency, which was also noted in other cases in which drip irrigation was used under similar management, i.e., high frequency pulses with limited amounts of water (Phene and Beale, 1979). When the amount of available water reservoir of the root is not much greater than that supplied by the daily irrigation, a fraction of that amount, if added over several days, might be influencial in changing soil water tension in the root zone and consequently growth responses. Such a system gives a powerful tool to control plant development.

The best agronomic performance was achieved in the treatment receiving the least irrigation. In this treatment the vegetative growth was strongly restricted (Tables 6 and 7), while the reproductive development was encouraged as characterised by yield, earliness (Table 8) and by the abscission rate (Table 9). It is unclear which biological system was affected by the specific root zone conditions and led to these phenomena. Carmi and Shalhevet (1983) attributed such effects to endogenous factors which were affected by the restricted growth of cotton roots. Their work was conducted in pots under field conditions and under nonlimiting fertigation management. In the present work, water stress conditions built up in plants receiving the smallest irrigation rates as indicated by low levels of leaf water potentials (Table 4). It is hard to distinguish under field conditions between possible effects of the specific shallow root system and soil water tension in the restricted root zone on the vegetative and reproductive responses of the plants. It is reasonable to assume that an interaction of both factors existed.

The irrigation management described in the present work was characterised by frequent irrigation pulses up to twice a day at mid-summer. Such a management may result under conditions of restricted root zone to cycles of dehydration and rewatering in the plants. The effect of such a situation on plant physiology may be different from that caused by elongated and stable stress conditions. It is possible that under such daily cycles, the vegetative organs will be more sensitive and inhibited in growth than the bolls. In such a case, a preferable reproductive development will occur as noted in the treatment which was the less irrigated and showed signs of water stress.

The concept of irrigation based on restricted and shallow root zone was developed for a double-cropping system in which the soil profile is dried by the winter crop. Potential advantages of raising cotton under such an irrigation management may be valid for single crop systems of early sown cotton, at least with regard to water use efficiency. It is questionable if the build up of a shallow root system is possible under conditions of a wet soil profile at the end of the winter period. It is possible that the roots will not be affected by the limited shallow wetting around the drippers, will grow to a depth and make the establishment of a shallow root zone impractical. Another possibility is that the roots will prefer to develop from their early stage of growth in the upper wetted zone as a result of a better aeration, nutrition and soil tillage. A later discontinuation of the humidity gradient in the soil, resulting from the roots activity in the upper layer, may create conditions in the root zone similar to those which characterise plants sown on a dried soil profile.

REFERENCES

Bielorai, H., Mantell, A. and Moreshet, S., (1983), 'Water relations of Cotton', in *Water Deficits and Plant Growth*, Academic Press, Inc., 49 - 87.

Bresler, E., (1977), 'Trickle-drip irrigation: principles and application to soil water management', *Adv. Agron.*, 29: 343 - 393.

Bunch, J.A., (1978), 'Effect of water stress on leaf expansion, net photosynthesis and vegetative growth of soybean and cotton', *Can. J. Bot.*, 56: 1492 - 1498.

Carmi, A., (1986), 'Cotton development as affected by root zone volume and plant density', *Field Crop Res.*, 13: 25 - 32.

Carmi, A. and Plaut, Z., (1982), 'A new agrotechnique for control of cotton growth and earliness of yield based upon drip irrigation system', *Proc. Belt. Cot. Prod. Res. Conf.*, Las Vegas, 58 - 59.

Carmi, A. and Shalhevet, Y., (1983), 'Root effects on cotton growth and yield', *Crop Sci.*, 23: 875 - 878.

Dasberg, S. and Bresler, E., (1985), *Drip Irrigation Manual*, Inter. Irrig. Inform. Center, Bet Dagan, Israel, 5 - 8.

Eaton, F.M. and Ergle, D.R., (1953), 'Relationships of seasonal trends in carbohydrate and nitrogen levels, and effect of girdling and spraying with sucrose and urea to the nutritional interpretation of boll shedding in cotton', *Plant Physiol.*, 28: 503 - 520.

Grimes, D.W., Dickens, W.L. and Yamada, H., (1978), 'Early season water management for cotton', *Agron. J.*, 70: 1009 - 1012.

Grimes, D.W., Dickens, W.L. and Anderson, W.D., (1969), 'Functions for cotton (*Gossypium hirsutum L.*) production from irrigation and nitrogen fertilization variables: II. Yield components and quality characteristics', *Agron. J.*, 61: 773 - 776.

Grimes, D.W., Anderson, W.D., Yamada, H. and Hughes, S.W., (1982), 'Irrigation effects on cotton biomass partitioning, *Cot. Prod. Res. Conf.*, Las Vegas, 51 - 52.

Jackson, E.B. and Tilt, P.A., (1968), 'Effects of irrigation intensity and nitrogen level on the performance of eight varieties of upland cotton, *Gossypium hirsutum L.*', *Agron. J.*, 60: 13 - 17.

Klepper, B., Taylor, H.M., Huck, M.J. and Fiscus, E.L., (1973), 'water relations and growth of cotton in drying soil', *Agron. J.*, 65: 307 - 310.

Phene, C.J. and Beale, D.W., (1979), 'Influence of twin-row spacing and nitrogen rates of high frequency trickle-irrigated sween corn', *Soil Sci. Soc. Am. J.*, 43: 1216 - 1221.

Willoughby, P. and Cockroft, B., (1974), 'Changes in root pattern of peach trees under trickle irrigation, *Proc. und Int. Drip Irrig. Congr.*, 239 - 442.

16 Control of tomato fruit yield and quality through fertigation

BNAYAHU BAR-YOSEF

INTRODUCTION

Irrigation and fertilisation are the most important management factors by which farmers control plant development and yield. The introduction of trickle (drip) irrigation and fertilisation (fertigation) (Goldberg *et al.*, 1976; Dasberg and Bresler, 1985) opened new possibilities for controlling water and nutrient supply to crops, and maintaining the desired concentration and distribution of ions and water in the soil. The improved control under drip fertigation is due to the simultaneous application of water and nutrients via the emitters (Bar-Yosef, 1977), the accuracy of water and fertiliser application and the ability to control irrigation frequency, which together with irrigation rate determine the soil-root volume of plants.

To optimise crop production, one needs to know two basic relationships: a) the consumption rate of nutrients as a function of time, which results in optimal fruit yield; b) the relationship between concentration of nutrients in the soil and soil solution within a given soil-root volume, and the rate of uptake of these elements by plants. This function is needed in order to create in the soil-root volume nutrient concentrations which sustain the optimal consumption rate by the plant as defined in (a).

The objective of this work was to study the effects of fertigation levels and nutrient concentration in the water on total and export quality fruit yield of glasshouse tomatoes, and determine the growth curve and nutrient consumption vs. time relationship which results in optimal tomato production.

optimal tomato production.

MATERIALS AND METHODS

The experiment was conducted in a Venlo-type glasshouse in Sde-Nitzan, located in the southern, arid part of Israel (34°E, 30°N). The tomato variety was Grenandier, introduced from Europe. Planting took place on 17 October, harvesting started on 15 February and terminated on 25 May. The soil was sandy, with 3% clay and a pH of 8.2. The all-season, time-averaged maximum, minimum and mean temperatures inside the glasshouse were 29, 12 and 18°C, respectively. The accumulated evaporation from a class A pan located outside the glasshouse was 850 mm. The plant stand was 28,000 per ha.

Treatments are summarised in Table 1.

Table 1

Summary of treatments

Treatment	*Water*		*Nitrogen*	
	*Crop coeff.**	*mm*	*mg N/L*	*kg N/ha*
C1W2	0.75	505	100	39
C1W3	1.60	1,200	100	91
C2W2	0.75	570	200	88
C2W3	1.60	1,240	200	180
C3W1	0.30	240	400	74
C3W2	0.75	490	400	146
C12W12	0.57	335	165	53

* Irrigation (mm)/evaporation from class A pan (mm) located outside the glasshouse.

Water and fertilisers were applied via a trickle irrigation system (emitter discharge rate 2L/h) with laterals positioned midway between double rows of plants. Fertigation was applied daily with a solution of fixed N:P:K ratio (6:1.1:5). Due to the constant ratio, solution concentration is expressed hereafter in terms of N concentration only.

Fruits were picked up selectively every two days and sorted out for local and export markets. Four times during the season fruits were tested for quality parameters after a storage period of three days at 18°C. The parameters were: fruit redness, determined by eye relative to a reference red color; fruit firmness, determined manually relative to a reference firm fruit; and blotchy ripening and fruit discolouration, determined visually. Fruits were grouped into those which passed the given quality standard, and those which did not.

Soil samples were taken seven times along the season at three distances (10, 25 and 45 cm) and five depths (10 cm increments to 50 cm) from plant stem. In these soil samples, soil electrical conductivity and NO_3-N concentration were determined in a 1:1

soil:water extract; and P and K in the soil were determined in a $NaHCO_3$ 0.5 *M* extract (Bar-Yosef and Akiri, 1978). Soil matric water potential was determined by tensiometers inserted at lateral distances of 10 and 25 cm and vertical distances of 20 and 40 cm from a trickler-source located near a plant stem. On 23 March, 10 x 10 x 10 cm soil cubes were excavated from the soil-root volume of plants in representative treatments (three replicates), and roots were separated from the soil by dry sieving. The above-ground part of plants was sampled seven times during the season and dry matter content and nutrient percentage in the shoots and fruits were determined by conventional techniques (e.g., Bar-Yosef, 1977). Leaf water potential was measured by a pressure bomb and leaf conductance to water vapor was measured by a stomatal diffusion porometer (Slavik, 1974).

RESULTS AND DISCUSSION

Nutrients and water in the soil

All-season temporal and spatial average tensiometer readings and nutrient concentrations in the soil within the soil-root volume are described for three representative treatments in Table 2.

Table 2

Effect of treatments on soil matric water potential, electrical conductivity (EC), soil-solution NO_3-N concentration, and $NaHCO_3$, soluble P and K in the soil.*

Variable	*C1W3*	*C2W2*	*C3W1*
Matric potential (-centibar)	7.0±3	15.0±5	30.0±10
EC (1:1 extract, dS/m)	0.5±0.3	1.2±1.0	1.5±1.0
P (mg P/kg soil, $NaHCO_3$)	28.0±35	35.0±26	42.0±25
K (mg K/kg soil, $NaHCO_3$)	210.0±45	265.0±55	480.0±230
NO_3-N in soil solution (mg N/L)	65.0±50	330.0±250	870.0±400

* Averaged over total growth season and sampled soil volume.
± indicates spatial and temporal deviation from the mean.

The analysed soil-root volume was a cylinder with a depth and radius of 50 and 45 cm, respectively, found previously to be characteristic of tomato plants grown under drip in a sandy soil (Bar-Yosef *et al.*, 1980).

The various fertigation combinations created a considerable range in average soil matric water potential, soil electrical conductivity and NO_3-N, P and K concentrations in the soil (Table 2). The large deviations from the mean results stem from fluctuations with time and distance from the point source.

Fruit yield

Total and export quality yields are summarised in Table 3. At fertilisation levels C1 and C2 (100 and 200 mg N/L), total yield (about 160 tons/ha) did not change when the irrigation rate was raised from 0.75 to 1.60 evaporation from a class a class A pan.

Table 3

Total and export quality fruit yield

Treatment	*Total*	*Yield (tons/ha) exportable*	*% export*
C1W2	165 a	69 b	42 b
C1W3	166 a	32 c	19 c
C2W2	153 a	83 a	54 b
C2W3	166 a	36 c	22 c
C3W1	63 c	46 c	73 a
C3W2	113 b	74 ab	65 ab
C12W12	100 b	76 a	76 a

Different letters in a given column indicate significant differences at the P = 0.05 level.

However, the elevated water rate reduced the export quality yield by a factor of 2, or more. Irrigation to evaporation ratios of 0.30 and 0.57 (C2W1, C12W12) apparently limited total fruit yield (63 and 100 tons/ha respectively), but, resulted in maximum percentage of export quality yield (73 and 76 %).

When increasing the compound-fertiliser concentration in the water at irrigation level W2 from 200 to 400 mg N/L, a significant reduction in total yield was obtained (113 vs. 165 tons/ha). The export quality yields was maximal at 200 mg N/L, minimal at 100 mg N/L, and not significantly different than the maximum at 400 mg N/L (Table 3). The percentage of export quality yield increased with decreasing water rate, or increasing fertiliser concentration in the water (Table 3).

Fruit quality parameters

Several fruit-quality parameters which determine adequacy for export, were investigated in relation to their dependency on irrigation and fertilisation rates (Table 4). Fruit firmness, redness and discolouration were not significantly affected by treatment. The incidence of blotchy ripening increased with enhanced irrigation rate, but was unaffected by elevated fertiliser concentration in the water.

Table 4

Effect of treatment on fruit quality parameters*

Treatment	*Fruits*			
	not red	*soft*	*blotchy*	*discoloured*
	(% of all fruits)			
C1W2	31	69	41b	59
C1W3	35	57	75a	66
C2W2	21	64	63b	46
C2W3	40	57	65ab	65
C3W1	14	53	19c	46
C3W2	24	72	36b	72
C12W12	27	72	32b	60
	NS	NS		NS

* After 3 days of storage at 18°C during March.
  Different letters in a given column indicate significant differences at the P = 0.05 level.
  NS not significantly different at the P = 0.05 level.

Correlations between fruit yield and quality, and system variables

To explain the above-mentioned effect of treatments on total (TY) and exportable (XPY) yields (Table 3), TY and XPY were correlated with water and nutrient input rates, on the one hand, and with average soil matric-water potential, EC and nutrient concentrations in the soil-root volume at the time of maximum fruit production rate, on the other hand (Table 5).

Table 5

Multiple regressons between yield (TY, XPY*) and input[a] and system[b] variables

Regression	R^2
$TY = 17.78 - 2.41\ (EC)^2$	0.94
$XPY = 1.05 + 15.43(CF) - 8.78(CF)^2$	0.94
$XPY = -0.78 + 14.48(CF) - 8.93(CF)^2 + 0.024(CN) - 0.00005(CN)^2$	0.99

For maximum XPY : (CF) = 0.81
(CN) = 240 mgN/L

* TY and XPY are accumulated total and export quality yields (kg/m^2 up to 24 April.
[a] CF = irrigation to evaporation (class A pan) ratio; CN = NO_3-N concentration in the irrigation water during the main growth season (mg/L).
[b] EC = soil electrical conductivity (dS/m) on 22 March.

The results indicate that TY was best correlated with the square of EC, which accounted for 94% of the total variablity in TY. The negative correlation between TY and EC indicates that under the experimental conditions, increasing EC was the main factor which limited total yield. It is noted that EC was controlled in the present study by the fertilisation level via the water (Table 2). Other causes of salinisation (e.g., Cl^-) could affect TY somewhat differently.

XPY was best correlated with the irrigation to evaporation (class A pan) ratio, which is sometimes called crop coefficient (CF) (R^2 = 0.94). The nitrogen concentration in the water (CN) and CF together accounted for 99% of the variability in XPY (Table 5). From the obtained relationship between XPY and CN and CF, one can calculate that in order to get maximum exportable yields CF and CN should be 0.80 and 240 mg N/L, respectively, during the main growth season. The CN value thus obtained is close to the NO_3 concentration in a full strength Hoagland solution, and somewhat lower than the gross-average NO_3-N concentration found in the soil solution of soil-root volumes of plants in treatment C2W2 (330 mg N/L), which gave the optimal fruit yield.

Fruit quality parameters were correlated with all available system variables (Table 6).

Table 6

Multiple regressions between fruit quality parameters and system variables*

Regression		R^2
[% blotchy]	= -31.0+8.34(TETA)	0.96
[% discoloured]	= 162.59-72.7(%P-fruits)+0.045(soil-K)+ +0.03(N-applied)	0.91
[% red]	= 63.73+28.22(EC)-0.072(soil-K)	0.91
[% firm]	= 77.86+0.011(NO_3-N in soil solution) -2.97(N-uptake)+8.56(P-uptake)	0.99

* Soil and plant parameters were determined on 27 March; quality parameters were averaged over the March harvest.
TETA (gravimetric soil water content, %), EC (soil electrical conductivity dS/m), NO_3-N (mg N/L soil solution) and K (mg/kg, extracted by $NaHCO_3$ 0.5 M) were averaged over total sampled soil volume.
Nutrient application and uptake rates are expressed in g/m^2.

Blotchy ripening was the only parameter which was highly correlated with a single variable, namely, the gravimetric soil water content (TETA). This agrees with our previous conclusion (Table 4) that blotchy ripening was significantly affected by irrigation rate. The incidence of discoloured and not fully red fruits was positively correlated with K concentration in the soil. The EC and NO_3 concentration in the soil solution were positively related to enhanced percentages of red and firm fruits, respectively. Elevated P-uptake and % P in fruits increased % firm fruits, and decreased % discoloured fruits, respectively. Accumulated N (mostly organic) in plants was negatively correlated with fruit firmness. Other factors related to nutrient supply and nutrient concentrations in plant tissue and in the soil were also searched for correlation with fruit quality parameters, but no consistent trend could be detected.

Crop-management related plant parameters

Some data on selected plant parameters pertinent to glasshouse tomato production are given in Table 7 for three representative treatments.

Table 7

Effect of treatment on some plant parameters

Parameter	*Treatment*		
	C1W3	*C2W2*	*C3W1*
Dry matter (tons/ha)	18.4 a	18.3 a	15.0 b
% dry matter in: shoots*	15.4 b	17.1 a	17.7 a
fruits[1]	5.1 c	7.0 b	10.0 a
Elements uptake (kg/ha): N	420 c	440 bc	480 a
P	109 a	100 a	70 b
K	960 a	944 a	782 b
Root system: (kg d.m/ha)	171 a	106 ab	162 a
depth,radius (cm)[2]	50,30 a	40,25 a	35,20 a
Leaf water potential (-bar)[3]	6.4 b	7.8 b	11.3 a
	10.5 c	12.6 b	15.5 a
Leaf conductance (mms^{-1})[4]	-	4.4 a	3.2 b

* On 19 February.
[1] On 28 March.
[2] 80% of the roots were found within this soil cylinder.
[3] At noon on 19 February and 28 March, respectively.
[4] On 19 February.
Treatments with different letters differed significantly at the P = 0.05 level.

Water stressed and excessively fertilised plants (C3W1) produced less dry matter, had a higher dry matter concentration in their shoots and fruits and took up more N than unstressed plants (C2W2, C1W3). Even though P and K concentrations in the soil in treatment C3W1 exceeded those in treatments C2W2 and C1W3 (Table 2), their cumulative uptake was smaller than in the latter treatments (Table 7), due to lower soil-water content in treatment C3W1. As expected, the uptake of those nutrients (P and K) whose mobility in the soil depend mainly on diffusion was more strongly affected by the reduced soil-water content than were nutrients which move mainly by mass flow (e.g. NO_3).

Leaf water potential in treatment C3W1 was lowest among the studied treatments, which explains the low leaf conductance in this treatment relative to C2W2. Root weight and soil-root volume in the various treatments did not differ significantly. However, a trend of decreasing soil-root volumes with declining irrigation rates was observed (depth , radius of 35 , 20 cm in treatment C2W1 vs. 50 , 30 cm in treatment C1W3, Table 7). The overall root dry matter weight comprised about 1% of the total cumulative canopy + fruit dry weight.

Plants which produced low total fruit yield had characteristic dry matter and N accumulation vs. time curves which remarkably differed from those characterising plants which have not been exposed to water stress or excessive fertilisation (Figures 1 and 2).

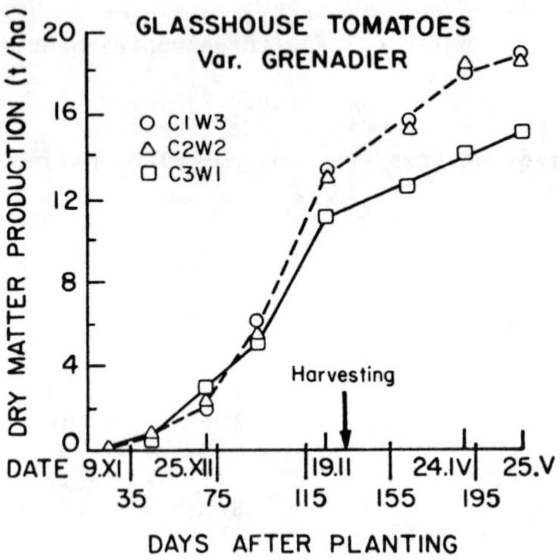

Figure 1 Dry matter accumulation as a function of time after planting. Symbols refer to three representative treatments

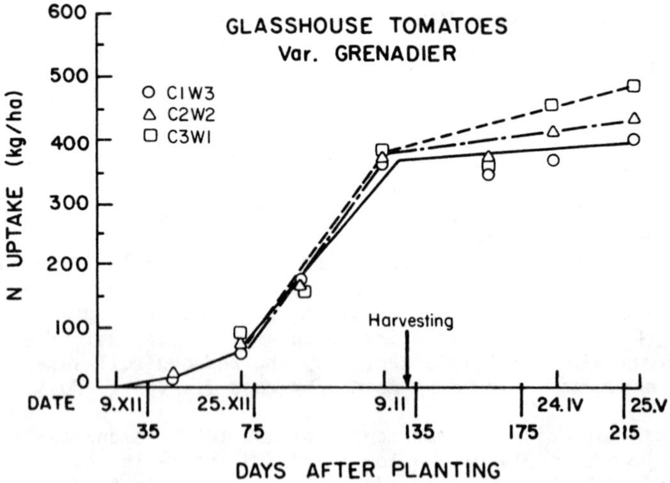

Figure 2 Reduced-N accumulation in plants as a function of time after planting. Symbols refer to three representative treatments

The characteristic curves of plants which gave optimal yield (C2W2) may serve as guidelines for monitoring plant development under climatic conditions similar to those reported herein, and for supplying N via the water along the growth season. Special attention should be paid to water and fertiliser application during the period starting about 100 days after planting, since by then most differences in dry matter production between treatments have been initiated.

During the first month of fruit harvesting (130-160 days after planting), nitrogen-uptake rate in all treatments was very small (Figure 2). This could stem from reduced root weight per plant at that time (Leonard and Head, 1958), due to competition between fruits, canopy and roots for the limited supply of carbohydrates. Following the month of low N consumption, the plants recovered and the N absorption rate increased in proportion to the N concentration in the irrigation solution. It yet needs to be studied whether foliar fertilisation during this critical growth stage may enhance plants productivity.

CONCLUDING REMARKS

As stated in the introduction, good fertigation management must be based on sound information on temporal nutrient and water consumption and dry matter production rate by plants, concentration of nutrients in the soil and irrigation water and soil-root volume of crops, which simultaneously lead to optimal plant growth and fruit yield. The presented results provide the information for a glasshouse tomato variety that was grown on a sandy soil, and enable one to analyse the factors that limit total fruit yield and fruit quality.

Total fruit yield was best (and negatively) correlated with soil electrical conductivity. Export quality fruit yield was best (and negatively) correlated with irrigation rate, due to the adverse effect of excessive soil water content on blotchy ripening of fruit. Optimal irrigation to evaporation (class A pan) ratio, and NO_3-N concentration in the irrigation water during the main growth season, should be 0.81 and 240 mg N/L, respectively.

REFERENCES

Bar-Yosef, B. and Akiri, B., (1978), 'Sodium bicarbonate extraction to estimate nitrogen, phosphorus and potassium availability in soils', *Soil Sci. Soc. Am. J.*, 42: 319 - 323.

Bar-Yosef, B., (1977), 'Trickle irrigation and fertilization of tomatoes in sand dunes: Water, N and P distributions in the soil and uptake by plants', *Agron. J.*, 69: 486 - 491.

Bar-Yosef, B., Stammer, C. and Sagiv, B., (1980), 'Growth of trickle irrigated tomato as related to rooting volume and uptake of N and water', *Agron. J.*, 72: 815 - 822.

Goldberg, D., Gornat, B. and Rimon, D., (1976), *Drip Irrigation - Principles, Design and Agricultural Practices*, Drip Irrigation Scientific Publications, Kfar Shmaryahu, Israel.

Dasberg, S. and Bresler, E., (1985), *Drip Irrigation Manual*, International Irrigation Information Center, Publ. No. 9, Bet Dagan, Israel.

Leonard, E.R. and Head, G.C., (1958), 'Technique and preliminary observations on growth of roots of glasshouse tomatoes in relation to that of the tops', *J. Hort. Sci.*, 33: 171 - 185.

Slavik, B., (1974), 'Methods of studying plant water relations', *Ecological Studies*, 9: 136.

17 Optimal management of growth in a greenhouse, with tomato seedlings as an example

IDO SEGINER

GENERAL

The usual objective of growers is to maximise their net profit from the agricultural processes they are setting in motion. The process of growing a crop may be viewed as a system with internal rules, inputs and outputs. The inputs are usually environmental conditions, the rules are mostly biological response functions to the environment, and the usual output is the yield. The overall problem of the grower is a decision problem: how to maximise his objective (profit) function, subject to various constraints, some imposed by the basic mechanisms of the system and some by other resource limitations.

A simple objective function, J, to be maximised, is

$$J = pY - C, \qquad (1)$$

where Y is yield, p is the price of the yield and C is the total cost of production. All these three factors are usually affected by the management of the system, M. More intensive management (or 'control') usually increases both cost and yield, while price may be affected by timing and quality, which, in turn, are also affected by management.

Consider now the limited problem of managing a greenhouse crop by controlling just the aerial environment (solar radiation, air temperature etc.). The system as a whole, consisting basically of the crop and the greenhouse around it, is affected by uncontrollable outside conditions, O (outside solar radiation, air temperature etc.). However, since some air-conditioning equipment (such as heater, fan and CO_2

generator) is available to impose a certain control, M, the crop itself may sense an inside environment, I, different from the outside environment. The state, S, of the crop, changes with time in response to the inside environment, which, in turn, is a function of the outside environment, and the control (decision, management) variable, M. Thus

$$dS/dt = g\{S,O,M\}, \tag{2}$$

where g is a function describing the dependence of the rate of change of S with time on S itself as well as on the outside weather and applied control. If the cost of operating the greenhouse at the level M is denoted by $c\{M\}$, an expression of the form

$$H\{M\} = i\ g\{S,O,M\} - c\{M\}, \tag{3}$$

known as the Hamiltonian, can be constructed. The multiplier i represents the intensity of cultivation (management), namely how much is the grower willing to pay for the marginal unit of production (it is also known as Lagrange multiplier or adjoint variable). At any point in time, the operating level, M, should be selected in such a way that the Hamiltonian, H, is maximised. To find the best choice of i for the crop, the crop development throughout the season is simulated for a range of values of i, utilising equations *(2)* and *(3)*. At the end of the simulated season, the objective (profit) function *(1)* can be evaluated, and the value of i for which J is maximal is then selected for the actual operation of the greenhouse.

NIGHT-TIME HEATING

Let us consider now two simplified examples of applying the method outlined above. The first relates to the night-time heating of tomato seedlings: Figure 1 shows the rate of growth of tomato seedlings as a function of night temperature for a fixed day-time temperature of 25°C.

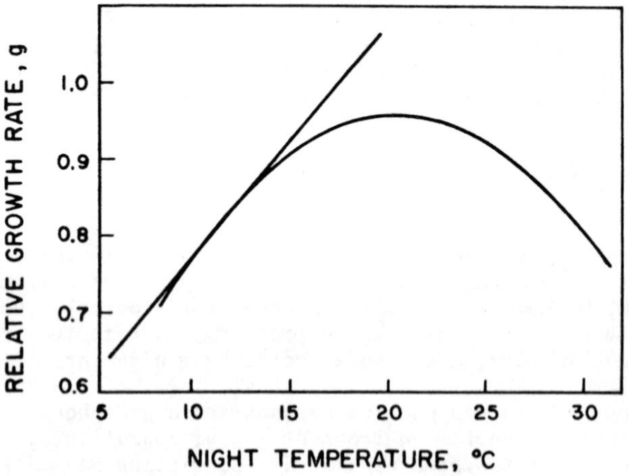

Figure 1 Rate of growth of tomato (cF. 104) seedlings as a function of night temperature for day temperature of 25°C. The tangent point indicates the optimal night temperature

The cost of maintaining a certain inside air temperature is approximated by

$$c\{M\} = k(T_a - T_o) \qquad (4)$$

where T_a and T_o are inside and outside air temperatures, respectively, and k is a constant. Thus, equation *(3)* becomes

$$H\{T_a\} = ig - k(T_a - T_o) \qquad (5)$$

which yields a maximum of $H\{T_a\}$ at

$$dg/dT_a = k/i \qquad (6)$$

as indicated in Figure 1.

The simplest objective function for the present case assumes pY to be constant. This is an acceptable simplification, provided that the crop is harvested at a predetermined size (not necessarily predetermined time). It must also be assumed that the price of the produce, p, remains constant with time. Under the assumption pY = const, J attains a maximum whenever the cost of production, C (equation *(1)*),

$$C\{T_a\} = [F + k(T_a - T_o)]\Theta \ , \qquad (7)$$

attains a minimum. In equation *(7)* F represents the fixed costs, such as rent, the next term in the brackets represents the cost of heating (equation *(4)*), and Θ is the time required to produce transplantable seedlings (termination time). The termination time can be approximated by

$$\Theta = Y/g \qquad (8)$$

where the yield Y is the relevant element of the state of the transplantable seedlings. The rate of growth, g, is assumed to depend only on T_a, which, in turn, is assumed to be constant for any choice of the intensity i.

Equation *(6)*, in conjunction with Figure 1, shows that higher values of i (higher intensity of management), imply higher values of T_a, as well as higher values of the growth rate g. These higher rates of growth result in a shorter termination time Θ. With these relationships in mind, equation *(7)* shows that as i increases, the fixed costs, $F\Theta$, decrease, while the heating costs probably increase. The optimum is found at some intermediate value of i, leading to an optimal inside temperature T_a.

The present model is simple enough for the optimal intensity to be found by minimising the objective function directly. Otherwise, equation *(2)* could be used to simulate the development of the seedlings for any given value of i, leading to the selection of its best value.

In a more detailed analysis of the heating problem (Seginer and Raviv, 1984), the optimal night temperature for tomato seedlings (cF. 104) was estimated at 11°C.

CO_2 ENRICHMENT

A second example of applying optimal management to tomato plants involves the enrichment of the greenhouse air by CO_2 (Seginer *et al.*, 1986). This application is more complex and requires a continuous adjustment of ventilation and CO_2 levels, according to the changing outside weather conditions. The optimal value of i was found by a series of simulations, which yielded the results of Figure 2.

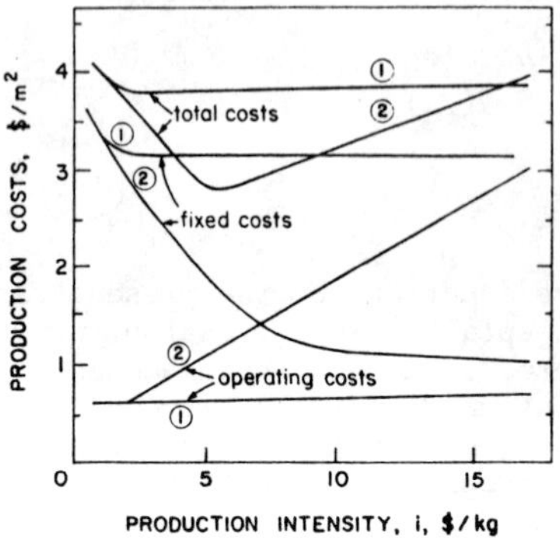

Figure 2 Operating costs, fixed costs and total costs of growing tomato seedlings in two greenhouses: (1) without CO_2 enrichment, and (2) with CO_2 enrichment. The total costs show a minimum at production (management) intensity i = 6 \$/kg

Two greenhouses are compared in Figure 2: Greenhouse 1, lacking a CO_2 generator, and Greenhouse 2, supplied with CO_2 at the rate required to minimise equation *(3)*. The figure shows clearly how the enrichment costs of Greenhouse 2 increase with the intensity of management, i. As a result, the rate of growth, g, increases, thus reducing the termination time, according to equation *(8)*. This, in turn, reduces the fixed costs. A minimum total cost of C = 2.8 $\$/m^2$ is achieved at i = 6 \$/kg.

In contrast to Greenhouse 2, the operator of Greenhouse 1, which lacks a CO_2 generator, cannot improve the inside conditions even when the value of i is high and 'allows' him to spend more on environmental improvement. Therefore, beyond an intensity of i = 3 \$/kg, not much is happening. The result is that the least total cost to produce the required yield is much higher than in Greenhouse 2, namely C = 3.8 \$/kg (rather than 2.8 \$/kg).

Once an optimal value of i is available, equation *(3)* can be minimised for any given outside environment O, and crop state S, to yield the recommended (optimal) operation level, M, of the air conditioning equipment. It seems that for relatively short periods of time (perhaps one month) the dependence on S may be ignored. As a result, a table can be constructed and stored in the greenhouse computer, which, for every set of outside conditions *(O)*, gives the recommended inside set-points *(M)*.

The recommended ventilation rate for tomatoes in the type 2 greenhouse (typical February in Israel) is presented in Figure 3, and the corresponding recommended CO_2 enrichment rate is presented in Figure 4.

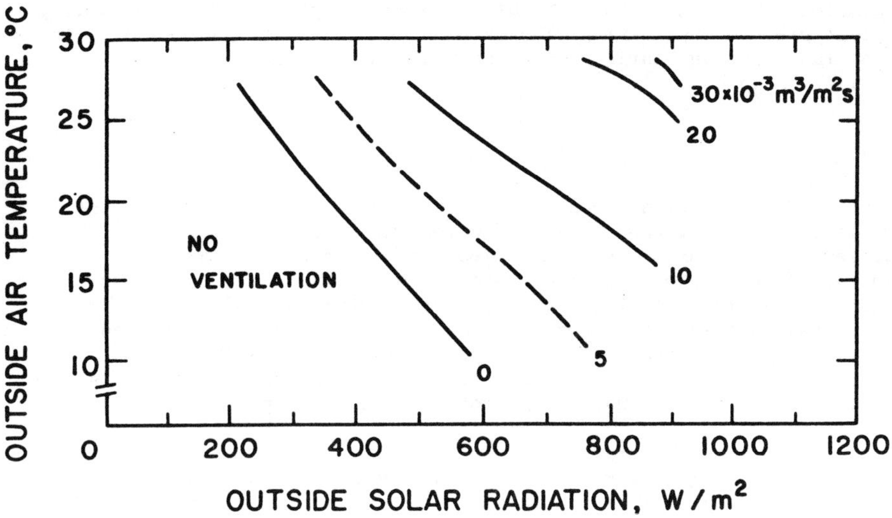

Figure 3 Recommended ventilation rates for the CO_2 - enriched Greenhouse 2, as a function of outside solar radiation and air temperature

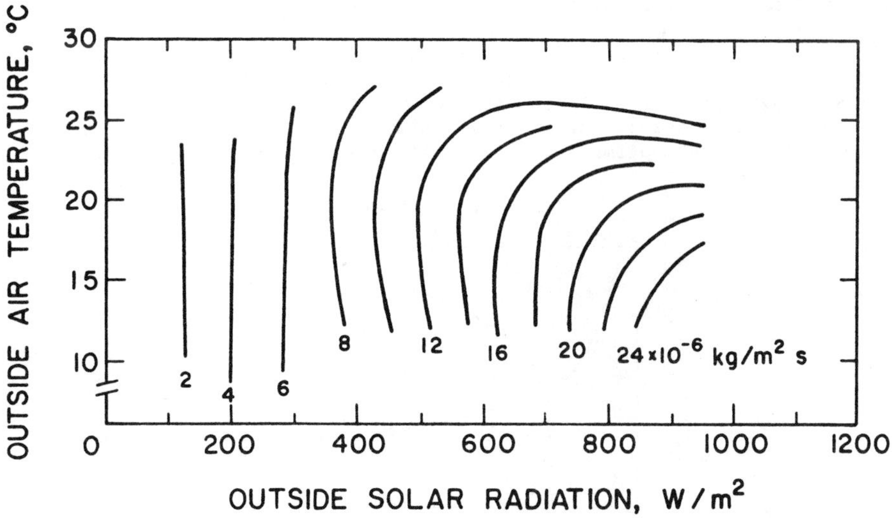

Figure 4 Recommended CO_2 - enrichment rate for Greenhouse 2, as a function of outside solar radiation and air temperature

The results show that the ventilation rate is kept to a minimum, namely just to prevent overheating. The rate of CO_2 enrichment increases with solar radiation and enrichment is maintained even under moderate ventilation rates.

VARIOUS PRODUCTION CONSTRAINTS

The objective function, equation *(1)*, was formulated to maximise the profit for a grower who has a production quota he may not exceed. Furthermore, the examples all assumed a fixed price for the yield. What happens if these two assumptions are no longer valid? Let us first consider the effect of removing the quota constraint. Since there is no unconstrained economical operation, there must be somewhere in the system another constraint. Several possible examples are: limited working hands, limited capital and limited greenhouse space. Let us consider the latter situation and formulate an appropriate objective function.

Intuitively, a grower who has limited greenhouse space, would strive to increase his profits by increasing the number of harvests (cycles) per year from the available area. To achieve this he would tend to increase his management intensity, i. The higher the market price for his produce, the higher his incentive to increase i. This is in contrast to the quota limiting case, where the market price does not affect the decision-making process.

The objective function for the space limiting case is

$$J\{i\} = [pY - C\{i\}]/\Theta\{i\}, \qquad (9)$$

where the total cost of producing a crop, C, as well as the time required to complete one growth cycle, Θ, are both functions of i. The functions $C\{i\}$ and $\Theta\{i\}$ for tomato seedlings in the CO_2 enriched greenhouse (Figure 2), can be used to demonstrate the effect of the market price p on the optimal selection of i in the present case (it is assumed, for simplicity, that the climatic conditions are stationary). Figure 5 shows $C\{i\}$ and $\Theta\{i\}$ as well as the optimal value of i for any given market price p.

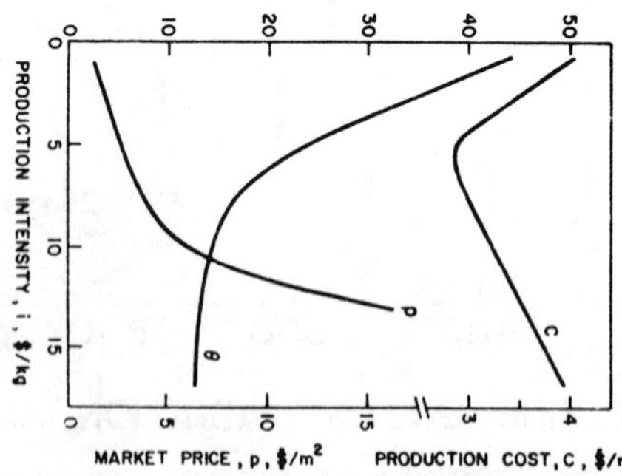

Figure 5 The dependence of the optimal management intensity, i, on the market price p, for a space-limited greenhouse operation

As expected, higher prices justify higher management intensities. The useful range of i is, however, limited, since the additional time saving at high management intensities is quite small.

If the market price, p, is a function of time, this can also be incorporated in the decision-making process. In this case p becomes a function of the termination time Θ, namely $p\{\Theta\}$, and since Θ is $\Theta\{i\}$, the price itself becomes a function of i. The objective function now becomes

$$J\{i\} = [p\{i\}Y - C\{i\}]/\Theta\{i\} \tag{10}$$

Let us consider, as an example, a case where the price increases with time. Intuitively, this should motivate the grower to delay his production, namely select a smaller value of i. The same result also emerges from the maximisation of equation *(10)* with respect to i, as is shown elsewhere (Seginer, 1987), together with other examples involving variable prices.

SUMMARY

Crops whose rate of growth is significantly affected by their environmental conditions (such as temperature and CO_2 concentration), can be manipulated by controlling their environment. In certain cases, a single-parameter optimal control (management) can be formulated. The optimal value of that parameter, i, is obtained by maximising an appropriate seasonal objective function J. Once the optimal value of i for a certain operation is found, tables can be constructed, which enable a greenhouse controlling computer to select the best inside conditions for any given set of outside conditions.

The selection of the objective function, J, depends on the economical constraints of the production system, and should be the first step in the grower's management decisions.

The weakest point of the method briefly described here, is the lack of detailed crop response functions, $g\{S,O,M\}$. An example has been worked out for tomato seedlings, based on data from the literature, to illustrate the method.

REFERENCES

The present paper is based on the following material :

Seginer, I. and Raviv, M., (1984), 'Optimal night temperatures for greenhouse tomato seedlings', *Sci. Hort.*, 23: 203 - 216.

Seginer, I., Angel, A., Gal, S. and Kantz, D., (1986), 'Optimal CO_2 enrichment strategy for greenhouses. A simulation study', *J. Agric. Eng. Res.*, 34: 285 - 304.

Seginer, I., (1987), 'Optimal greenhouse production under economic constraints', Manuscript.

PART V
MODELLING MANAGERIAL ACTIVITIES AND ECONOMIC ANALYSIS

18 Towards optimal pest control in cotton

DAN WALLCH, ELIEZER KLETTER, YAIR SACHS
AND EUGENE D. UNGAR

INTRODUCTION

Heliothis armigera is the first lepidopterous pest to appear on the cotton crop in Israel. It is most prevalent from June to mid or late July, and during this period destroys mostly flower buds and also some bolls. Control has been based on a threshold strategy suggested by the Ministry of Agriculture. Counts of the insect are carried out about twice a week in every field, and if the population of medium and large larvae exceeds the fixed threshold, the grower is supposed to spray.

It has been suggested that these recommendations are not fully consistent with the known behaviour of the plant and insect. In particular, the decision whether or not to spray should depend on total damage over the season, and not on the population at a single time (Wallach, 1980). Wallach also indicated that the weighting of medium versus large larvae was not reasonable.

It seems then advantageous to replace the empirical control strategy by a strategy based explicitly on current understanding of the cotton crop, the insect and their interactions, as embodied in computer simulation models. A number of cotton crop models have been developed that can be coupled with insect models (McKinion *et al.*, 1974; Wang *et al.*, 1977; Curry *et al.*, 1980; Wallach, 1980). These models have been used to examine the effects of various pest populations and of particular control strategies. The pests studied include *Lygus hesperus* (Gutierrez *et al.*, 1977), *Heliothis zea, H. virescens* and *Anthonomus grandis* (Brown *et al.*, 1979), *Anthonomus grandis* alone (Curry *et al.*, 1980) and *H. armigera* (Wallach, 1980).

Much less has been done on deriving optimal pest control strategies. Shoemaker (1981) has reviewed many of the applications of optimisation techniques to pest control. While these techniques can be used even for complex assumptions concerning insect dynamics, they generally require that the effect of the insect on the economic crop yield be quite simple. For cotton, however, this is certainly not the case. The cotton crop can compensate to a large extent for insect damage, so that yield loss is far from proportional to damage, and furthermore, the effect of damage varies with the timing of the removal of fruiting bodies, and with the sizes of fruiting bodies removed (Kletter and Wallach, 1982). As a result, in order to optimise pest control in cotton, both the crop and insect dynamics must be taken into account. This was done by Talpaz *et al.* (1978) in their study of boll weevil control. They calculated the optimal insecticide doses for each of 25 different four-day periods using a gradient-based method.

The optimisation technique of Talpaz *et al.* is quite general, but requires repeated evaluations of the overall cotton - boll weevil simulation model. Many of the cotton growers in Israel are linked to central computing facilities, but it is not realistic to envision very frequent runs of a cotton model on these computers. This paper suggests an algorithm for recommending *H. armigera* control decisions in cotton that uses an approximate optimisation procedure and requires only two evaluations of the cotton model. The algorithm is designed for real-time use. In particular, it is envisaged that the algorithm will be invoked after each new insect count becomes available, in each field. The algorithm then recommends whether or not to spray at that time.

THE COTTON MODEL

The cotton crop in the model is described by total plant weight, by the number of fruiting bodies of each age, and by the weight of fruiting bodies of each age. The equations of the model describe the time rates of change of these state variables, that is, the rate of total plant weight gain, the rates of bud production and bud and boll shedding, and the rate of fruiting body weight gain. These rates depend on the state of the plant. In particular, the basic feature of the model is the negative feedback of bolls present on the crop on bud production and bud and boll retention.

Environmental effects are included in the model only implicitly, through the parameter values. The model is then adapted to describing how differences in crop state (caused, for example, by different levels of insect damage) would affect subsequent crop behaviour in a particular environment. This environmental limitation is less restrictive than it might at first seem, since summer weather in Israel is fairly similar from year to year, and since a large part of the cotton crop is managed (and in particular is irrigated) according to a standard pattern.

The model used here is similar to that described in Wallach (1980). The basic difference is that the current model considers in considerably more detail the variation in crop-growth over the course of the growing season. In practice, this means that where a single parameter was used before to describe the average crop-growth response, there are now several parameters, one for each time period. The additional details are mainly based on a series of nine different bud and boll removal treatments carried out in 1978 (Kletter and Wallach, 1982) and on a series of 10 different damage treatments carried out in 1979 (Ungar *et al.*, 1987). Since the justification of the form of the model was largely presented by Wallach (1980), it will not be repeated here, but just a brief description of the model is presented below.

Time is measured in degree-days for cotton development, denoted by $^\circ D_{cot}$. The method of calculation is explained by Marani (1979). To reduce computing time, a time step of $70^\circ D_{cot}$ (roughly one week) is used in obtaining the solution. The model parameters therefore are defined for each $70^\circ D_{cot}$ period. The fruiting bodies are also aggregated into age classes that span $70^\circ D_{cot}$. Thus class 1 contains fruiting bodies aged 0 to $70^\circ D_{cot}$ at the end of each time step, etc.

Bud production is assumed to begin at $210^\circ D_{cot}$ and so the numerical solution is begun at this time. (Time step 1 then covers the period 210 - 280°Dcot, etc.). Buds are assumed to flower at the end of the fifth age class. Boll opening, on the other hand, does not occur at a fixed age, but rather at an age which decreases as the season progresses. Boll load, N, is defined as the number of bolls in age classes 7 through the last class of growing bolls.

Bud production is negatively related to boll load. Buds are first susceptible to shedding when in age class 1, and then again in age class 3. A constant 27% of the buds in age class 1 are marked for shedding early in the season, but this fraction increases to 100% by the end of the season. A constant 7% of the buds in age class 3 are marked for shedding over the period when boll load is zero. Thereafter, the number marked is an increasing function of boll load. Late in the season all buds are marked for shedding.

Small bolls (i.e., fruiting bodies in age class 6) can also be marked for shedding. Marking begins during time step 7, and the number marked increases with boll load.

The rate of weight gain of fruiting bodies in age class a during time step s, denoted $W_{a,s}$ (g/fruiting body/$70^\circ D_{cot}$), is fixed for the first six age classes:

$$W_{a,s} = 0.076 \,, \qquad a = 1,\ldots,4$$
$$W_{a,s} = 0.13 \,, \qquad a = 5,6 \qquad (1)$$

For older growing bolls

$$W_{a,s} = (W_t - W_y)/N \qquad a \geq 7 \qquad (2)$$

where W_t is the total plant weight gain and W_y is the weight gain by fruiting bodies in classes 1 - 6 during period s. The rate for older bolls is not allowed to exceed 1.2, or go below 0.57 g/$70^\circ D_{cot}$. The seed cotton per open boll, C, is calculated from dry matter, D, according to

$$C = -0.93 + 0.878D \qquad \text{g/boll} \qquad (3)$$

The effect of insect damage to fruiting bodies is simply to reduce the number of buds or bolls in the appropriate age class. If insect damage involves several age classes, the model distributes the damage in proportion to the number of fruiting bodies in each class. For the age classes susceptible to shedding, insect damage first simply reduces the number of fruiting bodies marked. Only if damage exceeds the number to be marked, is the number marked increased. Fruiting bodies once marked are assumed not to be attacked by insects.

THE *HELIOTHIS* MODEL

The model adopted for *H. armigera* is similar to that described in Wallach (1984). It describes a population with age structure, subjected to a natural mortality which varies between classes but is constant over time and for all ages within a class. The basic equation is

$$n_i(a,t) = n_i(a - \Delta a, t - \Delta t)\ \exp\ (-\mu_2 \cdot \Delta t) \qquad (4)$$

where $n_i(a,t)$ is the population density of age a in class i at time t, and μ_i is the mortality in class i. It is somewhat more convenient to deal with S_i, the probability that an individual survives through class i :

$$S_i = \exp(-\mu_i(h_i - l_i)) \qquad (5)$$

where l_i and h_i are, respectively, the lower and upper age limits for class i. Both age and time are measured in degree days for *Heliothis* development, denoted $D°_{hel}$, so that $\Delta a = \Delta t$ for all classes. The initial conditions for $n_i(a,t)$ are

$$n_i(a,t_b) = 0 \quad , \qquad i = 1,\ldots,4 \qquad (6)$$

where t_b is the time when recruitment (egg lay) begins. The boundary conditions are

$$n_i(l_i,t) = n_{i-1}(h_{i-1},t) \quad , \qquad i = 2,\ldots,4$$
$$n_1(0,t) = R(t) \qquad (7)$$

$R(t)$, the recruitment curve, is assumed to consist of two Gaussian curves separated in time:

$$R(t) = 0 \quad , \qquad t \leq t_b \qquad (8)$$
$$R(t) = A\{\exp[-(t-t_o)^2/(2\sigma^2)] + (1/b)\exp[-(t-t_o-g)^2/(2\sigma^2)]\},\ t_b < t$$

This is based on the observed general form of the population curves in the region of this study, and corresponds to two generations of moths.

The effect of insecticide is assumed to be an instantaneous mortality of some fraction of the individuals in each class. It is assumed that in field counts only a fraction of the individuals in class i are actually found, and so the counted populations are adjusted according to the estimated 'scouting efficiency' for each age class.

In order to make the model sensitive to the specific conditions of each field, certain of the model parameters are recalculated from all the available insect counts for the field, each time the algorithm is invoked. This is equivalent to assuming that the model parameters are fixed for each field but are imperfectly known, and that improved estimates are obtained as more data become available. The parameters that are recalculated are t_b = the time at which recruitment begins, A = the multiplicative factor in the recruitment function, $\theta_1 = \theta_2 =$

the probabilities (assumed equal) that eggs and small larvae are counted, and t_o = the time of the first peak of the recruitment curve. Late in the season, when the population results mainly from the second recruitment peak, the parameter b, which is the ratio of the heights of the first and second recruitment peaks, is also calculated. The start of recruitment, t_b, is calculated independently of the remaining parameters, based on the earliest times when individuals are observed in classes 3 and 4. The parameter t_o, and the parameter b late in season, are calculated by a nonlinear least squares minimisation routine. At each iteration, A is chosen so that the calculated and observed populations summed over counting dates, and summed as well over classes 3 and 4, are equal. Then θ_1, (and therefore θ_2) is chosen so that the calculated and observed populations are equal for classes 1 and 2 summed together and over counting dates.

Other combinations of parameters to be calculated could have been chosen. However, the choice made here ensures that the model reproduces the essential aspects of the field counts, while keeping the calculations to a minimum. Given the large amount of scatter in the data, it did not seem worthwhile to try to fit the data more closely by adjusting additional parameters. It is also because of this scatter that it was chosen to combine classes or dates in certain calculations, in an empirical attempt to smooth the data somewhat.

The *Heliothis* model serves two essential functions. First of all, it provides an estimate of the current population which is based on smoothing all the available data, rather than on simply the most recent counts. It thus makes decisions less sensitive to jumps in the observed population levels. The current population level determines the amount by which an insecticide application on the current day can reduce further damage. Secondly, the model predicts the future population and therefore future damage. This also affects the decision whether or not to spray, because the response of a cotton crop to damage depends on the overall damage pattern.

THE OPTIMIZATION TECHNIQUE

The objective function to be minimised here is $F(S)$, the total cost associated with control strategy S. Explicitly

$$F(S) = C_s n(S) + C_y [Y_o - Y(S)] \qquad (9)$$

where C_s and C_y are respectively the costs of a single spray application and a unit of yield, $n(S)$ is the number of spray applications in strategy S, Y_o is the yield in the absence of damage and $Y(S)$ is the yield for strategy S. Thus the first term in equation *(9)* is the cost of control measures, and the second the cost associated with yield loss.

It is assumed that the field can be sprayed at the current time (i.e., at the time when the question whether or not to spray is posed), and then at subsequent intervals of 40°D_{hel} (about four or five calendar days) up to some cutoff date. The set of allowable strategies then consists of all possible combinations of spray and no spray decisions over all desicion times. The spray dosage is assumed fixed. For 10 decision times, for example, there are $2^{10} = 1,024$ possible strategies.

The choice of a $40°D_{hel}$ interval between decisions means that the optimal spray date may be missed by up to about two days. Since the insect population does not seem to have a very sharp peak, this does not seem a serious problem. The chosen interval also automatically eliminates the possibility of sprays spaced more closely than $40°D_{hel}$, but it seems unlikely that such strategies are desirable given that, according to the insect model, the *Heliothis* population is still seriously depressed $40°D_{hel}$ after a spray.

Given the number of possible strategies considered here (in the order of 10^6 early in the season, then less as the season advances), an exhaustive examination of them all using the combined cotton and *Heliothis* models is not feasible. The technique proposed here is to use the insect model alone to choose a small number of candidate strategies, and then to use the cotton model to choose among those.

The two qualities sought for the candidate strategies are firstly that they contain as close an approximation as possible to the true optimal strategy, and secondly that they can be generated using an efficient algorithm.

The set of candidate strategies proposed is the set of strategies that minimise total damage by *H. armigera* for a given number of spray applications. Initially, that is, an approximate objective function $F'(S)$ is defined as

$$F'(S) = C_s n(S) + C_d D(S) = C_s[n(S) + rD(S)] \qquad (10)$$

where C_d is the cost of one unit of damage and $D(S)$ is the amount of damage (i.e., the number of damaged fruiting bodies) which results from strategy S. $F'(S)$ depends on the insect population, and so can be minimised using the insect model alone.

For different values of the ratio $r = C_d/C_s$ in equation *(10)* optimal strategies are generated with different numbers of spray applications. The candidate strategy with i sprays is denoted S'_i. As explained below, the two strategies of interest are that with the greatest number of sprays which does not include the date in question, and that with the least number of sprays which does include that date. In order to search efficiently for those strategies, it is helpful to have upper and lower bounds for the r values that will generate them. Let RMAX and RMIN be respectively the greatest and smallest possible reductions in damage that can result from a spray today. Then $r \leq 1/\text{RMAX}$ must lead to a spray on the current day and $r \geq 1/\text{RMIN}$ cannot lead to a spray. This establishes the limits on r within which the strategies required will be generated. RMAX and RMIN are easily calculated. RMAX is the reduction in damage due to adding a spray at the current time when there are no spray applications in the future, and RMIN is obtained when there are spray applications at every future date.

If cotton yield varied linearly with total damage, $F'(S)$ would be equivalent to $F(S)$, and the optimal strategy for the appropriate value of r in equation *(10)* would be the true optimal strategy (denoted S^*). Going a step further, if cotton yield decreased monotonically as total damage increased, but nonlinearly, then the overall approach here would still provide the true optimal strategy. In this case S^* cannot be determined directly using $F'(S)$ because the proper value r for equation *(10)* is unknown a priori. However, whatever the number of spray applications in S^*, say j, S^* must be the strategy with minimum total damage for that number of sprays, and therefore $S_j = S^*$. Then S^* will be selected when the strategies in $\{S'_i\}$ are compared using the objective function $F(S)$.

The response of cotton yield to damage. However, is more complicated still. It is known to depend not only on total damage, but also on the timing of damage, and on the distribution of damage among fruiting bodies of different sizes (Kletter and Wallach, 1982). In this case the set $\{S'_i\}$ need not include S^*. For example, among two strategies with the same number of spray applications, that with greater total damage but to smaller fruiting bodies might be superior, but will not be included in $\{S'_i\}$. It is difficult to estimate how serious a problem this represents (i.e., how different the selected strategy will be from S^*). It should be noted, however, that different strategies are unlikely to lead to extreme differences in damage pattern.

Regardless of the strategy, damage must be concentrated fairly early in the season, when *Heliothis* is present, and involve mostly buds, which are the preferred food of all larval size classes. The above arguments suggest that the set of candidate strategies $\{S'_i\}$ should contain a reasonable approximation to the true optimal strategy S^*. The derivation of the $\{S'_i\}$ can be efficiently carried out using the approach suggested by Birley (1979). Birley noted that when the recruitment curve is assumed fixed, as is the case here, the insect population at a given time will vary only as a result of different insecticide application patterns prior to that time. For the specific conditions considered here,insecticide affects only larvae, and not eggs. The youngest larvae, aged 30°D_{hel}, disappear from the population after at most 130°D_{hel}. Thus only insecticide applications during the preceding 130°D_{hel} affect the population at time t. During this period there are three decision times (assuming that t is itself a decision time), and $2^3 = 8$ possible combinations of decisions. There are therefore 8 possible population levels at any decision time t, and these are the states for the dynamic programming calculation.

For each state at time t_1, the control decision that minimises $F'(S)$ is determined. For any earlier time t_i and state k, the dynamic programming recursive equation states that the optimal control decision is that which minimises the sum of the total cost up to the next decision time t_{i-1}, plus the minimal total cost from the resulting state at t_{i-1} to the end of the season. The amount of damage in each time interval is calculated from the insect model.

The calculation thus proceeds backwards in time, until at some time the states depend on decisions earlier than the current time. Those decisions are automatically set to no spray, and the true spray dates are taken into account. When the current date is reached then only a single state is considered, and the optimal strategy for that state is the sought for candidate strategy.

In general the proposed approach requires that once found, all the candidate strategies, up to some reasonable upper limit on the number of spray applications, be examined using the cotton model. However, given certain features which are observed for the particular problem considered here, it can be shown that it is sufficient to consider just two of the candidate strategies. This simplification is of considerable practical interest, since execution of the cotton model is a time-consuming portion of the algorithm.

This simplification depends on the fact that the real interest is not in the full optimal strategy, but only in knowing whether or not that strategy includes a spray application on the current date. Future strategy is important in that it alters the effect of a spray at the current time but decisions here are never made in advance. Rather, at each new date the algorithm is rerun to determine the recommended strategy at that time.

The first observed feature of interest here is that cotton yield declines more and more rapidly as total damage increases, i.e.,

$$dY/dD < 0 \qquad d^2Y/dD^2 < 0 \qquad (11)$$

The observed general behaviour follows straightforwardly from the cotton model. For low levels of damage the crop compensates by reducing natural shedding, and the resulting effect on yield is slight. Eventually however, as damage increases, the crop can no longer fully compensate and yield begins to decrease.

It is also observed that for given insect parameters total damage for strategy S'_i decreases, but more and more slowly, as i increases. That is

$$\Delta D_i = D(S'_{i+1}) - D(S'_i) < 0$$
$$\Delta^2 D_i = D(S'_{i+2}) - 2D(S'_{i+1}) + D(S'_i) > 0 \qquad (12)$$

This behaviour is logical, since the strategies S'_i with successively larger i are generated by increasing r, which is equivalent to lowering the reduction in damage required before a spray is worthwhile.

The third observed feature of interest is that if S'_i includes a spray at the current time t_n, then all S'_j for $j > i$ also include a spray at t_n. This is not the case for decision dates other than t_n. The damage reduction due to adding a spray application at t_i depends on the control decisions at other times in the interval $[t_i - 130°D_{hel}, t_i + 130°D_{hel}]$. Spray applications within that interval reduce the effectiveness of a spray at t_i. Thus S'_j relative to S'_i ($j > i$) may add a spray in the interval around t_i, and as a consequence drop the spray at t_i. The decision time t_n however has a special status, in that different strategies cannot have different decisions at earlier times. It is indeed possible to prove, for the particular case of S'_1, that if S'_1 generated for some value r_1 of r, includes a spray at t_n, then so do all S'_j for $j > 1$. The proof involves showing that no strategy without a spray at t_n can be optimal for $r' > r$, because the same strategy with the first spray date, say t_i, replaced by t_n would be superior. This follows from the fact that a spray at t_n alone must result in less damage than a spray at t_i alone, and furthermore the remaining sprays in the proposed strategy will be as effective, or more effective, if the initial spray is removed from t_i to t_n. However, no general proof that the observed behaviour must always occur has been found.

It is now possible to show that it is worthwhile to spray at the current time if and only if

$$Y(S'_k) - Y(S'_{k-1}) > C_s/C_y \qquad (13)$$

where S'_k is the candidate strategy with the smallest number of sprays that includes a spray at the current time. Inequalities *(11)* and *(12)* imply that yield increases, but more and more slowly, as the number of sprays i in S'_i increases. The first term in the objective function $F(S)$ (equation *(9)*) on the other hand increases linearly with the number of sprays. As a result $F(S'_i)$ has a unique minimum at some value i^* of i. If inequality *(13)* holds then $F(S'_k) < F(S'_{k-1})$, implying that $k < i^*$. Therefore S'_{i^*} includes a spray application at the current time, according to the third observed feature of the results indicated above. If inequality *(13)* does not hold, then $i^* > k$ and i^* does not contain a spray at the current date. Finally, then, the cotton model need only be run with the two candidate strategies S'_k and S'_{k-1}.

The above discussion assumes that there is a fixed, predefined harvest date. It is worthwhile to consider how to relax this restriction, since damage to reproductive bodies in cotton is known to delay yield buildup (Kletter and Wallach, 1982). Thus different harvest dates may be appropriate for crops with different levels of damage. Given the costs associated with each harvest date, it is possible to determine the adjusted yield value for each strategy, defined as the maximum over harvest dates of yield value minus harvest costs. (The cost associated with the harvest date increases as harvesting is delayed, essentially because the crop must then be protected for a longer time from late season insect pests, and because the probability that rainfall will cause yield losses increases). If it is now assumed that there is a diminishing return in terms of adjusted yield from adding spray applications, then equation *(13)* is still the necessary condition for spraying, if the yield values are replaced by adjusted yield values. It should be noted that in practice it may be quite difficult to evaluate the costs associated with different harvest dates. In particular, these costs may depend on the harvest schedule of the whole farm.

FIELD TESTS

The optimisation technique described above was applied to historical data from six commercial fields in the coastal plains region of Israel, and control recommendations were generated for each date on which insects were counted. This indicates the expected behaviour of the algorithm for real, and therefore 'noisy' data. In generating the candidate strategies S'_i, an upper limit of 0.1 was generally set for r in equation *(10)*. If the candidate strategy for this value of r did not include a spray application at the current date (i.e., if a spray application at the current date reduced total damage by less than $1/r$ = 10 fruiting bodies), it was assumed that a spray at the current date was not worthwhile. Indeed, in all those cases where the full algorithm was used, and where S'_k was generated using $r \geq 0.1$, the recommendation was not to spray at the current time.

Yields at 1,680°D_{cot} are used in equation *(13)*. The possibility of delaying harvesting is not considered. The ratio C_s/C_y is taken as 1/6, based on 1983 prices. That is, the cost of an aerial spray application over a field is approximately equal to the price of an additional 6 g seed cotton/m^2 over the field. It may be noted that for the cases considered here this represents roughly 1% of total yield, while the standard error in yield measurements is generally about 5% of the mean. The fact that the yield loss that justifies control cannot be measured is one of the major difficulties in determining an optimal control strategy for *H. armigera* empirically.

For fields 1 and 2, no candidate strategy generated with $r \leq 0.1$ included a spray at the current date. In these fields then no insecticide application against *H. armigera* is recommended. This agrees with the strategy actually followed in these fields.

For field 3, the algorithm recommended a control application on 14 June (since $Y(S'_k) - Y(S'_{k-1}) > 6$ for that date). The field was not sprayed at that time, and so the algorithm again recommended a control application on 17 June. Thereafter, no control applications are recommended. Roughly, then, there was a period of a few days in mid-june during which a spray application was worthwhile. If the field is not sprayed then it is no longer worthwhile to spray at later times.

Field 4 is adjacent to field 3 and was sprayed on 12 June. The results for this field thus indicate what the subsequent recommendations of the algorithm would have been, if the first spray recommendation had been followed. For field 4 after the spray application, no candidate strategy generated with $r \leq 0.1$ included a spray at the current time.

The recommendations of the algorithm then, if followed, would have led to a single early spray in these fields. This is nearly identical to the Israel Ministry of Agriculture recommendations, which led to a single spray on 12 June. On the basis of the model calculations, the yield increase in field 4 resulting from an early spray should be about 10 g/m^2. This then should be the approximate yield difference between fields 3 and 4. As already mentioned, such a difference will not be significant, given the experimental techniques available. The observed yields in fields 3 and 4 were 548 and 559 g/m^2, respectively. This at least does not contradict the results of the calculations.

For adjacent fields 5 and 6, the algorithm repeated the recommendation that insecticide should be applied at each date early in the season. Beyond approximately mid-June, however, a spray application is no longer worthwhile. Field 5 was sprayed on 10 June, more or less in keeping with the recommendations of the algorithm. No subsequent spray applications are recommended. The Ministry of Agriculture recommendations on the other hand led to three spray applications in field 5. There were no spray applications in field 6. Despite that, final yield did not differ significantly from that of field 5.

CONCLUSIONS

The cotton and insect models adopted here represent great simplifications of the true plant and insect dynamics. The approximations made are, however, quite different in the two cases. The cotton model assumes that in the absence of damage the crop develops according to a fixed pattern. In fact, for the purposes of insect control, it is only really important that the effect of damage be identical among crops, which is a less stringent condition. While even the latter condition is certainly not met in general, it may be a reasonable first approximation for a good part of the Israeli cotton crop. Indeed, the model has been found to perform reasonably well for a variety of damage patterns in a number of different cotton fields in Israel. It should be noted that the optimisation technique adopted here could be used for any other cotton model as well.

The insect model accepts that the population dynamics differ between fields, but assumes that within each field the insect parameters are fixed. That is, the model ignores the variability in survival and reproduction caused by changes in weather, population density, food availability or natural enemy populations. While these factors are known to be important (the role of natural enemies of *H. armigera*, for example, has been investigated by Bar *et al.*, 1979), it seems impossible at the present time to accurately describe all these effects on *H. armigera* population levels. The simple model used here seems capable of at least approximate predictions of future damage levels, and permits the use of an efficient dynamic programming algorithm for determining control recommendations.

Finally, the optimisation technique employed is itself only approximate. It has been argued, however, that it may reasonably be supposed to furnish recommendations similar to those of the true optimal strategy. It is probably impossible, given the experimental errors in yield measurements, to determine experimentally whether the present algorithm gives strategies superior to other methods. Certain advantages of the algorithm over empirical strategies are however clear. Firstly, the algorithm aids in understanding the basis for *Heliothis* control, since it is based explicitly on descriptions of the plant and insect. Secondly, the algorithm can be progressively improved, since it is based on a series of explicit assumptions, each of which can be tested and improved independently. Finally, changes in the system parameters (especially the cost of insecticide, or the price of cotton) can easily be incorporated in the algorithm.

The algorithm has been incorporated into a package of programs for cotton crop management (Wallach *et al.*, 1980) which is widely used on-line by the growers in the coastal plains region of Israel. The package already includes a program for input of insect counts, and so those are available to the algorithm.

Future work will concentrate on incorporating *H. armigera* control strategy into an overall strategy of cotton crop management, including irrigation and control of other insect pests. Since the cotton model is currently parameterised for Acala cotton, reparameterisation will be required before growers can use the full optimisation package for the Pima variety. This is important in view of the recent trend of an increasing proportion of the cotton acreage being sown to the Pima variety.

REFERENCES

Bar, D., Gerling, D. and Rossier, Y., (1979), 'Bionomics of the principal natural enemies attacking *Heliothis armigera* in Israel', *Environ. Entomol.,* 8: 468 - 474.

Birley, M.H., (1979), 'The theoretical control of seasonal pests - a single species model', *Math. Biosci.,* 43: 141 - 157.

Brown, L.G., McClendon, R.W. and Jones, J.W., (1979), 'Computer simulation of the interaction between the cotton crop and insect pests', *Trans. ASAE,* 22: 771 - 774.

Curry, G.L., Sharpe, P.J.H. and DeMichele, D.W., (1980), 'Towards a management model of the cotton - boll weevil ecosystem', *J. Environ. Mgt.,* 11: 187 - 223.

Gutierrez, A.P., Leigh, T.F., Wang, Y. and Cave, R.D., (1977), 'An analysis of cotton production in California: *Lygus hesperus (Heteroptera: Miridae)* injury - an evaluation', *Can. Ent.,* 109: 1375 - 1386.

Kletter, E. and Wallach, D., (1982), 'Effects of fruiting form removal on cotton reproductive development', *Field Crops Res.,* 5: 69-84.

Marani, A., (1979), 'Growth rate of cotton bolls and their components', *Field Crops Res.,* 2: 169 - 175.

McKinion, J.M., Baker, D.N., Hesketh, J.D. and Jones, J.W., (1974), 'SIMCOT II: A simulation of cotton growth and yield', *Computer Simulation of a Cotton Production System: User's Manual,* ARS-S-52, USDA, 27 - 82.

Shoemaker, C.A., (1981), 'Applications of dynamic programming and other optimization methods in pest management', *IEEE Trans. Automat. Contr.,* AC-26: 1125 - 1132.

Talpaz, H., Curry, G.L., Sharpe, P.J.H., DeMichele, D.W. and Frisbie, R.E., (1978), 'Optimal pesticide application for controlling the boll weevil on cotton', *Amer. J. Agr. Econ.,* 60: 469 - 475.

Ungar, E.D., Wallach, D. and Kletter, E., (1987), 'Cotton to bud and boll removal', *Agron. J.,* 79: 491 - 497.

Wallach, D., (1980), 'An empirical mathematical model of a cotton crop subjected to damage', *Field Crops Res.,* 3: 7 - 25.

Wallach, D., Kletter, E. and Dauvergne, M., (1980), 'Mathematical models for cotton crop management', in: Yaron, D. and Tapiero, C. (eds.), *Operations Research in Agriculture and Water Resources,* North-Holland, Amsterdam, 249 - 258.

Wallach, D., (1984), 'An algorithm for the short-term prediction of insect populations', *Prot. Ecol.,* 6: 181 - 197.

Wang, Y., Gutierrez, A.P., Oster, G. and Daxl, R., (1977), 'A population model for plant growth and development: coupling cotton-herbivore interaction', *Can. Ent.,* 109: 1359 - 1374.

19 Scheduling of cotton harvesting operations at the farm level

HOVAV TALPAZ, ELIEZER KLETTER AND EUGENE D. UNGAR

INTRODUCTION

Harvesting is a crucial operation of the cotton production process. Timing and scheduling of the many fields upto the harvest operation is important to achieve economic success, especially for large multifield cotton farms. In this study, a strategy is developed to account for plant characteristics in planning the harvest scheme subject to the environmental conditions. In particular, the climatic conditions of the Israeli coastal plains are considered.

Over recent years, experimentation at the Gan HaShomrom Experimental Station has shown that the cotton plant is capable of a full-yield compensation to quite high levels of direct bud damage by insect pests or bud shedding induced by water stress (Kletter and Wallach, 1982; Wallach and Kletter, 1984). These findings are of great significance to the grower since, in general, the same yield can be obtained with less intensive inputs of insecticides and greater flexibility in irrigation scheduling. However, full-yield compensation is invariably accompanied by a significant delay in yield accumulation. In the coastal plains region of Israel, a relatively undamaged crop reaches full maturity (100% boll opening) in the last week of September or the first week of October. The problem is that a delay in harvesting brings the grower into a period of rapidly increasing probability of rainfall.

To date, the grower has been highly apprehensive of the risk of yield and quality reduction that is assumed to result from exposure to rain accompanied by high winds.

In considering a pest control and irrigation strategy for a single field only, three alternatives need to be evaluated:

1. Low-risk timing of harvesting facilitated by standard management practices that avoid a delay in yield accumulation but entails a higher ratio of input cost per unit yield.
2. Delayed harvesting to permit full yield compensation resulting from reduced use of expensive inputs such as insecticide spraying. This entails taking the risk of rainfall damage.
3. Double harvesting to reduce the magnitude of possible rainfall damage. Inputs are reduced and yield accumulation delayed but a large portion of yield is harvested during a period of negligible rainfall probability and only the late-maturing fraction of yield is harvested during the relatively high risk period. This option has the following disadvantages:
   a. Extra harvesting costs are incurred.
   b. There may be yield loss from mechanical damage to unopened bolls at the first harvest.
   c. There may be a significant loss of lint quality at the second harvest.
   d. Since complete defoliation is required before harvesting operations can proceed, defoliation must commence at a slightly lower percentage boll opening than normally practiced. This may cause a reduction in yield. An example of such losses is shown in Figure 1, which depicts losses as sampled from 181 fields in the coastal plain regions of Israel. Figure 2 indicates the change in harvest losses resulting from rainfall intensity in the same area.

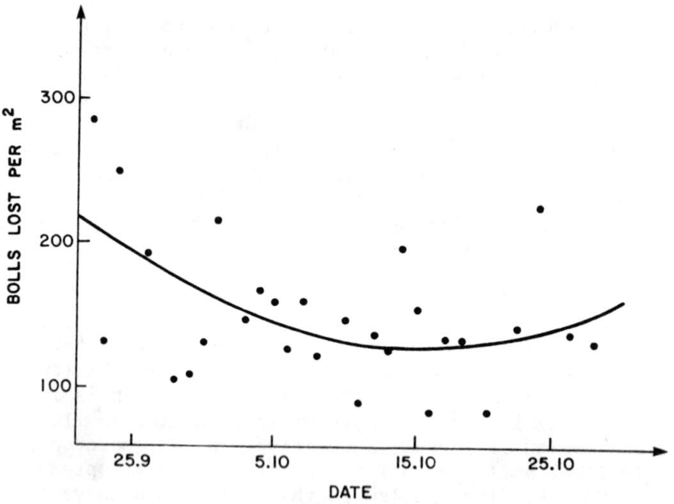

Figure 1 No. of seeds in yield loss per square meter as a function of harvesting date, 1982

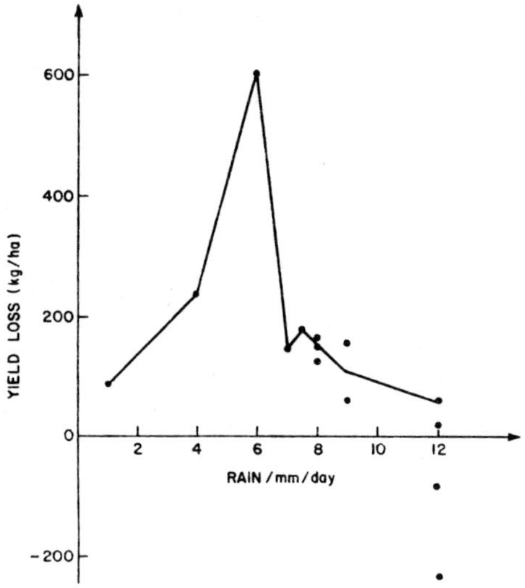

Figure 2 Seed difference in harvest losses per square meter as a function of daily amount of rain. The graph represents the average loss for each amount of rain

If the problem is extended to a whole-farm context, it is necessary to consider the effect of a delay in a particular field on the harvest scheduling of the farm as a whole. In some instances, a delay in one field may be unimportant if that field would have become harvestable together with a number of other fields. Since the harvesting capacity of the farm is generally a limiting factor, a rearrangement of the harvesting sequence of these fields may suffice. On the other hand, a delay in a small number of fields may cause a redistribution of harvesting load through time that cannot be met by the machinery and manpower of the farm without continuing harvest into a high-risk period. This creates the additional options of hiring extra harvesting capacity and of advanced defoliation and double harvesting of some fields. Here, the advantage of double harvesting derives not only from reducing the potential magnitude of rain damage but also from the fact that the need to hire extra harvesting capacity is reduced. Clearly, the number of possible harvesting schedules makes rational decision-making a formidable task. It was therefore decided to develop a computer algorithm that would generate a set of viable scheduling alternatives, together with an economic evaluation of the associated costs and benefits, from which the manager can select. Ultimately, such an algorithm would be linked to existing pest control recommendation programs and provide an evaluation of alternative strategies in terms of their impact on the farm as a whole.

The development of a harvest schedule algorithm requires quantitative information concerning:

1. The expected rainfall as a function of date.
2. The effect of harvest date and rainfall on harvesting loss and quality.

3. The effect of timing of defoliation and double harvesting on cotton yield, quality and harvesting losses.

This report summarises results obtained thus far in these three areas and presents a brief description of the harvest scheduling algorighm.

ESTIMATING THE EXPECTATION OF RAINFALL AS A FUNCTION OF DATE

Daily rainfall data for the 40-year period 1940 to 1979 were taken for three locations in the coastal plains region of Israel - Givat-Haim, Ein HaHoresh and Gan Shmuel. Rainfall at these locations was measured using a small-orifice rain gauge. All analyses carried out for the period September to November and data from the three locations were analysed as one data set. Two analyses were carried out, the first to establish the expectation of a rainfall event on a given day and the second to estimate the quantity of rain that would be obtained at a rain event.

1. *Expectation of a rain event*

For each date in the period September to November, the number of rain events in the data set was counted and expressed as a fraction of the total number of rain events (see Table 1).

Table 1

Sample probability of rainy days over specified periods (40-year average, all locations)

Period	*Rainfall class*	*Sample probability*
	Mean annual (mm/day)	0.013
	0 < rain < = 5	0.013
16/9 - 30/9	5 < rain < = 15	0.000
	15 < rain < = 30	0.000
	rain > 30	0.000
	Mean annual (mm/day)	0.075
	0 < rain < = 5	0.045
1/10 - 15/10	5 < rain < = 15	0.021
	15 < rain < = 30	0.007
	rain > 30	0.002
	Mean annual (mm/day)	0.125
	0 < rain < = 5	0.065
16/10 - 30/10	5 < rain < = 15	0.037
	15 < rain < = 30	0.014
	rain > 30	0.009

These figures were converted to a cumulative probability over the period and an exponential function fitted to the data. The differential of this function represents the expectation of a rain event as a function of the number of days from 1 September.

2. *Expected rainfall during a rain event*

Preliminary analysis of the data showed cumulative rainfall to increase exponentially with time, hence the following function was fitted :

$$R = a\Theta^{bt} \qquad (1)$$

where R is cumulative rainfall from 1 September (mm), a and b are coefficients to be estimated. Dividing R and a in equation *(1)* by the cumulative rainfall on 30 November of each year yields :

$$\rho = \alpha\Theta^{bt} \qquad (2)$$

where ρ is the proportion of total rainfall for the period September to November obtained up to time t, and α and a are coefficients.

Differentiating equation *(2)* with respect to t provides the proportionate instantaneous contribution of any point in time to total rain :

$$\rho' = b\alpha\Theta^{bt} \qquad (3)$$

Finally, the proportionate contribution to total rain of a given day (1 September $\leq t \leq$ 30 November) is given by the integral :

$$R_t = \int_t^{t+1} b\alpha\Theta^{bt}dt \qquad (4)$$

It is reasonable to assume that the extent of harvesting loss caused by rainfall is related to the intensity and duration of the rain event. Furthermore, on the basis of field experience of extension personnel in the region, wind intensity during and immediately after a rain event is also an important factor determining the extent of rain damage. Wind and rainfall intensity data are not available over a sufficient period of time to be used as a basis for the present analysis. Instead, the wind factor is ignored at this stage and the rainfall components of intensity and duration are treated together as daily rainfall. Four rainfall intervals are arbitrarily defined : up to 5 mm/day; 5 - 15 mm/day; 15 - 20 mm/day; over 30 mm/day.

Parameter estimation of equation *(1)* was carried out separately for the 2nd, 3rd and 4th rain intervals and the results are shown in Table 2. An excellent fit was obtained for rain intervals 2 and 3, though less so for rain interval 4.

Table 2

Estimation of parameters a and b in equation *(1)* for various rainfall intervals

Rain interval	*Parameter*				R^2
	a	t_a*	b	t_b*	
2	0.552	15.3	0.043	52.9	0.988
3	0.338	7.3	0.049	30.6	0.968
4	0.237	1.54	0.057	7.23	0.665

* t = statistic for the parameters a and b, respectively.

Sets of rainfall patterns can be stochastically generated using the above analysis. For a given day, a random number between zero and one is generated from a uniform distribution function. A rain event occurs if this number is greater than the proportionate expectation for rain on that day. The quantity of rainfall for the rain event is then defined by a variant of equation *(4)*, namely, where all rain intensities are lumped together. Another way of evaluating the effects of rain is to consider a few typical rain regimes. This is very important since an 'expected' rain schedule could distort, rather significantly, the effect of rain due to nonlinearities of the associated response. To do so, the 41 years of rain statistics have been categorised into 10 different patterns arbitrarily chosen. These are listed in Table 3.

Table 3

Categories of rainfall according to patterns and levels

Category number	*Rainfall quantity (mm in Sept - Oct)*	*Pattern description*	*% of occurrence*
1	0 - 6	-	26.8
2	> 6 - 15	dispersed	9.8
3	> 6 - 15	concentrated	17.1
4	>15 - 30	dispersed	14.6
5	>15 - 30	concentrated	2.4
6	>30 - 50	after Oct. 25	2.4
7	>50 - 60	dispersed	12.2
8	>50 - 60	concentrated	4.9
9	approximately 80	dispersed	4.9
10	approximately 80	concentrated	4.9

Each year was classified into one of these categories. The user of the harvesting algorithm (see below) may select any category by taking a particular year. Weighting of each solution may be done using the percent of occurrence.

EFFECT OF HARVEST DATE AND RAINFALL ON HARVESTING LOSS: DISCUSSION

The trend of an initial declining phase in the loss function with time can be explained in terms of more complete opening, and hence harvesting, of bolls with time. Later in the season, the trend for harvesting loss to increase can be explained in terms of a certain degree of natural falling of fully open bolls which may be aggravated by the mechanical effects of the harvesting machinery passing along the rows.

Based on measurements taken by extension personnel over a number of growing seasons, it was found that harvesting-loss of cotton containing 100 seeds is equivalent to approximately 170 kg seed cotton yield per ha.

On the basis of the combined 1982 and 1983 analysis, a 30-day difference in harvesting date would result in a difference of 90 kg seed cotton per ha according to linear regression analysis. The minimum of the polynominal expression occurs on 23 October and represents a reduction of 130 kg seed cotton loss per ha from the predicted value on 23 September. These results indicate that the reduction in harvesting loss with time is too small an effect to merit consideration in deciding whether to delay harvest of a particular field by a few days. A time effect on harvesting loss has not been incorporated into the harvest scheduling algorithm. This algorithm fits the risk averter decision-makers, since we feel that a few more years of losses due to rain - wind combinations should be studied before more affirmative conclusions are drawn. However, the gambler can use this algorithm by changing the loss parameters.

Quantification of the effect of rainfall on harvesting loss clearly requires many years of data and only provisional estimates can be provided at present.

On the basis of results from 1983, the scheduling algorithm assumes rain of up to 10 mm/day to have no effect on loss. (The mean increase of 16 seeds/m-row represents about 0.5% of yield. Barker (1982) found that rainfall of less than 14.5 mm/day had no effect on preharvest losses in the hill areas of Mississippi). For rain events of more than 10 mm/day, there is assumed to be a yield loss of 6 kg seed cotton per ha per mm rain. This value is based on field measurements in October 1977, when a rainfall of about 80 mm resulted in a loss of approximately 500 kg seed cotton per ha. Rain damage parameters in the scheduling algorithm will be refined as further data become available.

EFFECT OF DEFOLIATION TIMING AND DOUBLE HARVESTING ON COTTON YIELD AND QUALITY: RESULTS AND DISCUSSION

Percentage boll opening curves and timing of defoliation are shown for three locations in Figures 3 - 5.

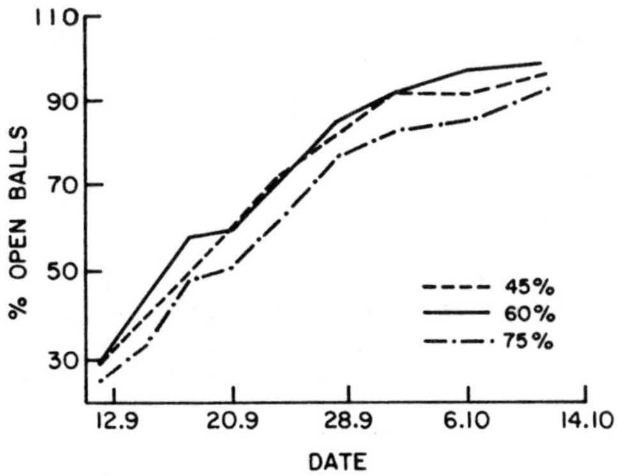

Figure 3 % of open bolls as a function of the date at Metzer. The arrows indicate the defoliation dates

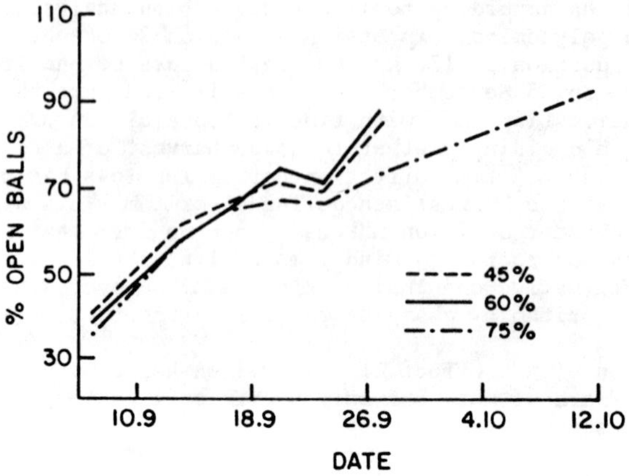

Figure 4 % of open bolls as a function of the date at Nachsholim. The arrows indicate the defoliation dates

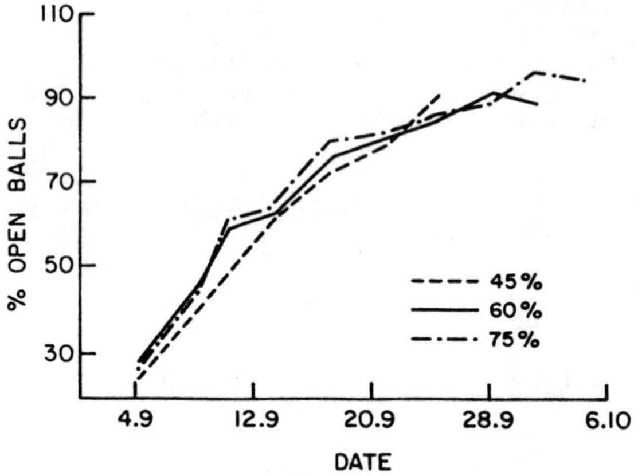

Figure 5 % of open bolls as a function of the date at Lehavot Haviva. The arrows indicate the defoliation dates

Earlier defoliation appears to accelerate boll opening. Advancing defoliation from 75 to 45% boll opening results in a 5-to-10-day advance in reaching 100% boll opening. On the whole, the rate of boll opening ranges between 2 and 3 per day. Results of yields and quality are given in Tables 4 - 7. Total hand harvest yield at Metzer was not affected by early defoliation. At Lehavot Haviva, total lint yield was significantly reduced by early defoliation and double harvesting by machine. At Nachsholim, defoliation timing had no effect on seed cotton or lint yield. Defoliation timing had a significant effect on mechanical damage at first harvest at both Lehavot Haviva and Nachsholim (Table 8).

Table 4

The effect of defoliation timing on cotton yield at Metzer

Treat.	*Seed cotton yield (kg/m-row)*						
	1/10	*6/10*	*16/10*	*Harv.I**	*Harv. II*	*Total*	*% lint*
45%	439	27	21	439	48 a	487	38.7
60%	446	31	24	477	24 b	501	38.7
75%	411	31	28	470	-	470	38.5

* Harvest I is defined as yield up to 17 days after first defoliation, remaining yield is classed as Harvest II. Numbers within a column followed by the same letter are not significantly different according to Duncan's Multiple Range Test.

Table 5

The effect of defoliation timing on lint quality at Metzer

Treatment	*Presley*	*Micronaire*	*Fibrograph*	*Uniformity*
	Harvested at 1/10			
45%	92.6	44.2	119.4	49.2
60%	92.8	44.6	122.0	49.8
75%	93.6	43.8	122.4	50.4
	Harvested at 6/10 and 16/10			
45%	93.0	37.6	122.2	47.8
60%	91.8	35.4	119.8	48.6
75%	91.8	37.2	118.8	49.0

Quality characteristics were not found to be significantly different on the 0.05 level according to Duncan's Multiple Range Test.

Table 6

The effect of defoliation timing on cotton yield at Lehavot Haviva and Nachsholim

Treat.	*Seed cotton (kg/m-row)*			*Lint yield (kg/m-row)*			*% lint*	
	Harv. I	*Harv. II*	*Total*	*Harv. I*	*Harv. II*	*Total*	*Harv. I*	*Harv. II*
Lehavot Haviva								
45%	407 b	27 a	435 a	134 b	8 a	142 b	32.9 b	
60%	418 ab	26 a	443 a	137 b	7 a	144 b	32.8 c	28.0
75%	446 a	17 a	463 a	152 a	5 a	156 a	34.0 a	
Nachsholim								
45%	505 a	79 a	584 a	164 a	24 a	188 a	32.5 b	
60%	523 a	62 a	585 a	171 a	19 a	190 a	32.7 a	30.6
75%	542 a	56 a	599 a	160 a	17 a	177 a	29.5 c	

Numbers within a column for a given location followed by the same letter are not significantly different at the 0.05 level according to Duncan's Multiple Range Test.

Table 7

The effect of defoliation on lint quality at Lehavot Haviva and Nachsholim

Treat.	*Presley*	*Micro-naire*	*Fibro-graph*	*Unifor-mity*	*Quality grading* *Commer-cial*	*Relative ranking*	*Colour*
Lehavot Haviva - 1st harvest							
45%	91	40	112	45	5.0	6.2	fair
60%	91	40	112	45	5.2	5.3	good
75%	91	40	112	45	4.8	6.5	good
2nd harvest							
	89	28	105	42	8.0	9.0	
Nachsholim - 1st harvest							
45%	92	40	111	46	6.5	6.8	fair
60%	92	40	110	45	6.4	6.3	
75%	91	40	111	45	6.2	6.6	
2nd harvest							
	92	31	109	44	8.0	9.0	

Table 8

The effect of defoliation timing on mechanical harvesting damage to unopened bolls at Lehavot Haviva and Nachsholim

Treatment	*Number of unopened bolls per 10 m-row* *Before harvest*	*After harvest I*	*Difference**
Lehavot Haviva			
45%	70.4 a	53.2 a	17.3 a
60%	54.5 a	43.6 a	10.9 ab
75%	39.3 a	35.1 a	4.3 b
Nachsholim			
45%	161.5 a	129.5 a	32.0 ab
60%	126.5 b	89.5 b	37.0 a
75%	107.0 b	93.0 b	14.0 b

* To adapt numbers to Table 8, a weighted average should be used. Numbers within a column for a given location followed by the same letter are not significantly different at the 0.05 level according to Duncan's Multiple Range Test.

Since the number of replicates was low, data for Lehavot Haviva and Nachsholim were analysed together (Table 9). This analysis indicates the significant effect of early spraying and double harvesting on seed cotton yield. It might be expected that defoliation as early as 45% boll opening would effect cotton yield through reduced boll weight resulting from an early reduction in the supply of photosynthates. The fact that only a small effect was obtained can be explained in view of the following:

1. Bolls cease to increase in weight about five days before complete opening.
2. Defoliation causes leaf shedding approximately five days after application. Thus, at an average rate of boll opening of 2.5% day, one can roughly state that defoliation at 45% is equivalent to a reduction in effective photosynthate supply from 70% boll opening.
3. Defoliation is effected in layers; the first application at 45% causes shedding of the uppermost level of leaf canopy only (though these leaves are youngest and most active).
4. The photosynthetic efficiency of leaves declines steadily towards the end of the season and a significant portion of boll weight increase derives from carbohydrate extracted from leaves and steams. Hence, defoliation as early as 45% boll opening, at worst, reduces photosynthate supply to 30% of the total boll number for a short period towards the end of their development.

Table 9

The effect of defoliation timing on cotton yield and mechanical harvesting damage for combined Lehavot Haviva and Nachsholim data

Treat.	*Seed cotton yield (kg/m-row)*			*Mechanical damage*
	Harvest I	*Harvest II*	*Total*	*Bolls/10 m-row*
45%	440 b	45 a	484 b	22.2 a
60%	453 ab	38 ab	491 ab	19.6 a
75%	478 a	30 b	508 a	7.5 b

* Defined as the difference between the number of bolls before and after harvest I. Numbers within a column followed by the same letter are not significantly different at the 0.05 level according to Duncan's Multiple Range Test.

It should be noted that the cotton in the field at Nachsholim was a slowly developing crop. This is evident from Figure 4 and from Table 10 which shows that a relatively high percentage of total yield was harvested at second harvest. Such a field is quite uncommon but was chosen for this experiment in order to examine a wide range of conditions. Defoliation at 75% followed by 1st harvest 17 days later still left over 9% of yield in the field. This illustrates the importance of a dynamic approach to defoliation and harvest scheduling that considers the *rate* at which bolls are expected to open. A computer program for predicting boll opening on the basis of field counts of boll numbers is available to the growers.

There was a consistent effect of double harvesting on micronaire and colour grading. Double harvesting results in a lower colour grade for 2nd harvest cotton, presumably due to discolouration from leaf material and dirt. Micronaire is reduced by about 10 points at 2nd harvest.

Table 10

The effect of defoliation timing on the allocation of cotton yield between harvests

Treat.	*% total seed cotton yield harvested at*						*Total seed cotton harvested as % of harv. I at 75% defol.*		
	Harvest I			*Harvest II*					
	L	*N*	*L+N*	*L*	*N*	*L+N*			
45%	93.8	86.5	90.7	6.2	13.5	9.3	97.5	107.7	101.3
60%	94.1	89.4	92.3	5.9	10.6	7.7	99.3	107.9	102.7
75%	96.3	90.6	94.1	3.7	9.4	5.9	-	-	-

L = Lehavot Haviva; N = Nachsholim; L + N = Combined Lehavot Haviva and Nachsholim.

CONCLUSIONS

Commencement of defoliation as early as 45% boll opening causes a small reduction in total cotton yield. However, additional financial losses incurred to the grower through mechanical damage to unopened bolls at first harvest and through a reduction in the lint quality of second-harvest cotton. Until more data becomes available from further experiments, the following parameters are taken in the scheduling algorithm. Defoliation is not permitted earlier than 50% boll opening.

Double harvesting is carried out for fields defoliated between 50% and 66% boll opening. This upper limit for double harvesting can be computed from knowledge of the expected cotton yield, lint percentage and economic parameters. Such a calculation will be used in the algorithm at a future stage of development. On the basis of results from the more typical Lehavot Haviva field, the algorithm assumes the allocation of yield between the first and second harvests to be 94% and 6%, respectively. It is estimated that the quantity of cotton left in the field by single harvesting (after defoliation at 75% boll opening) is approximately equal to any yield reduction by earlier defoliation and mechanical damage caused by double harvesting. Hence no yield reduction factor is taken for double-harvested fields. Percentage lint for second-harvest cotton is taken to be 93% of the first-harvest value. A reduction of two units in fibrograph and three units in quality grade is taken for second-harvest lint. In the scheduling algorithm, these effects are incorporated directly in the price of lint.

APPENDIX A

Algorithm for the scheduling of cotton harvesting operations

The algorithm accepts any number of user-determined harvesting scenarios and calculates for each a feasible schedule and its profitability. A harvesting scenario is defined by:

1. Whether or not double harvesting is allowed.
2. Whether or not the hiring of extra harvesting capacity is to be

considered.

3. If the hiring of extra harvesting capacity is permitted, whether one or two machines can be hired at once and over which period hiring may be considered.

The user may choose the default set of scenarios listed in Table 11 which covers the range of practice.

Table 11

The default set of scenarios for cotton harvesting

Option number	*No. of hired harvestors*	*No. of harvests*	*Minimum of % open bolls at 1st defoliation*
0	0	1	66
1	1	1	66
2	2	1	66
3	0	2	50
4	1	2	50
5	2	2	50
6	0	2	55
7	1	2	55
8	2	2	55
9	0	2	60
10	1	2	60
11	2	2	60

For each scenario, the algorithm computes the recommended harvesting date and defoliation dates for every field, the total number of days required to complete harvesting and the period during which extra harvesting capacity should be hired (if such an option is allowed). In addition, the algorithm provides economic summary estimations including the gross and net income, harvesting transportation, ginning and defoliation costs and the yield lost through rain damage. For a given scenario, a feasible schedule (taking into consideration holidays and weekends) is generated for any number of rainfall patterns. These rainfall patterns can be taken from historical data or generated stochastically using the analysis of rainfall data described earlier. Yield losses are deducted from an average seed cotton yield value provided by the extension officer. The program allows for quality effects of rain though no losses are incurred at present due to lack of data. The grower is advised to take into consideration not only the net profit, but also the robustness of alternative of different rainfall patterns.

The algorithm ranks all scenarios according to the net income of the whole cotton farm, and prints it out. The grower may select the 'best' (highest net income) or any other one based on his risk aversion behaviour. Information regarding the size and location of cotton fields within the farm, sowing date and variety sown are all drawn directly from the existing data base at the computer installation. The percentage boll opening curve for each field can be computed according to a standard set of parameters. Parameter values are based on observations from many fields over a number of seasons. Values are currently: for sowing on 1 April, boll opening commences on 15 August for Acala cotton and on 23 August or Pima cotton. Commencement of boll opening is advanced one day for every four days advancement in sowing before 1 April and delayed one day for every two days delay in sowing after 1 April (all varieties). Delay rate of boll opening is taken to be 3% up to 45% boll opening, 2.3% from 45 - 66% boll opening and 2% from 66 - 100% boll opening in Acala cotton, and 1.7% from 66 - 100% boll opening in Pima cotton. The program also accepts field counts of percentage boll opening and these figures override predicted default values up to the specified date. There is also an option of using an existing algorithm for the prediction of boll opening that provides an accurate prediction based on field counts of large bolls (Wallach and Kletter, 1981).

Default values for daily harvesting capacity and all input and output prices are available to the algorithm, but the grower can substitute values for most of these. The grower can specify days (weekends, holidays etc.) on which no defoliation or harvesting operations are to be scheduled. Computation of a feasible harvesting schedule is subject to a large number of constraints. The more important of these are:

1. For single harvest fields, the earliest feasible harvesting date is defined by 100% boll opening.
2. First, second and third defoliation sprays are applied 17, 11 and five days prior to the recommended harvesting date.
3. Defoliation cannot commence before 50% boll opening is reached.
4. Double harvesting is not carried out for fields defoliated after 66% boll opening.
5. There is a minimum daily acreage below which defoliation spraying cannot be ordered.
6. Pima variety is not double harvested.
7. Priority of fields with identical boll opening percentage on a given day is according to variety (Pima over SJ-2) and proximity of fields.
8. Start of harvesting operations is delayed to eliminate idle days.
9. Rainfall causes a delay in harvesting according to the quantity of rain.

It is intended to use the scheduling algorithm to determine the economic feasibility of joint acquisition of additional harvesting equipment by groups of growers. The program allows for the easy updating of all economic and biological parameters as new information becomes available.

REFERENCES

Barker, G.L., (1982), 'Equation for estimating cotton preharvest losses', *ASAE Trans.*, 271 - 276.

Kletter, E. and Wallach, D., (1982), 'Effects of fruiting form removal on cotton reproductive development', *Field Crops Res.*, 5: 69 - 84.

Wallach, D. and Kletter, E., (1981), 'Prediction of boll opening in a cotton crop', *Agron. J.*, 73: 763 - 767.

Wallach, D. and Kletter, E., (1984), 'Cotton crop response to bud and boll removal' (in press).

20 Economic analysis of cotton irrigation technologies

DAN RYMON AND GIDEON FISHELSON

INTRODUCTION

One might define the issue of optimal yield management as that of choosing among various alternatives of technology or combinations of technologies. In a static world this decision is relatively simple; assuming that all the available technologies are known, there are no elements of uncertainty and no effects of risk aversion. The issue becomes complicated once new technologies are (continuously) entering the feasible set. A common characteristic of a new alternative is the lack of information. With the acquiring of experience the lack of information gradually decreases, until the new alternative becomes a well known technology. Hence until then, almost by definition, elements of uncertainty are present and the attitude toward risk becomes an important factor in the decision of choosing technologies. Other issues that are related to new technologies involve special outstanding input requirements. Thus, relative high prices of inputs, shortages of capital and credit restrictions might result in nonadoption of a new technology although the calculations (made under simplifying assumptions) justify a full speed ahead for its adoption [1].

This study limits the choice of technology to irrigation technologies; the crop chosen to demonstrate the issue is cotton. The reason for the choice of cotton is its large share in the irrigated field crops in Israel (see Table 1 in Fishelson and Rymon, (1986)) as well as in several other countries. [2] The same approach holds, however, under completely different crop mixes, because the focus is on the issue of optimal irrigation technologies in general.

The issue of choosing the optimal irrigation technology is dealt with only rarely in the economic literature, while in practice it is a daily issue (some studies are listed below). The reasons we can see for this phenomenon are:

1. Lack of data while theoretical arguments and discussions are of no specific interest or use.
2. The nonfamilarity of many economists with the micro-data and the potential technologies.
3. The belief that the issue is taken care of by salesmen, extension services, advisers and the ear-to-mouth process of spreading innovative information in a conventional agro-economic environment.
4. There are only few organisations interested in this field. It falls in an interface between the (macro) investment in water resources (dams, drilling, conduits, reservoirs etc.) and the (micro) on-farm management and decision-making with regard to existing irrigation systems.

Yet every economist will agree that the appearance of new or modified technologies deserve systematic analysis and evaluation.

ECONOMIC CONSIDERATIONS IN CHOOSING AN IRRIGATION TECHNOLOGY

Disregarding uncertainty, the objective of minimising the total cost of production per unit output is obvious. In order to generalise the choice, the minimisation has to consider all the components that make up these costs. This breakdown of the total costs into detailed components is of significance since the technologies to be compared might vary significantly with respect to the contributions of some components.

Our study is an empirical one. The reader is saved from a discussion on theoretical cost analysis, and is led directly into the construction of isoquants (an 'isoquant' is defined as the locus of combinations of inputs that yield the same output level). For simplicity and convenience of graphing in two dimensions, the various irrigation costs were aggregated into two categories of production factors and their corresponding costs - costs of capital and costs of 'labour'; the latter includes the expense of tractors, fuel and miscellaneous costs, usually of minimal magnitude. It means, that the production function borne in mind might be defined as "output generated by the employment of 'labour' and capital". We distinguish between the two factors although they are both measured in monetary terms because of their different nature, i.e., they can be regarded as two different inputs. The costs that we consider are limited to those related directly to irrigation or those which are affected by the choice of irrigation technology. Considerations related to the human factor (need of skilled worker, easiness of work, etc.) and difficulties associated with after-sale-service of equipment do not limit, under Israeli conditions, the adoption of one technology over another. Secondary effects of the technology choice were also omitted, unless they are characterised later in the efficacy of land utilisation, in the other irrigation expenses or in the yield (quantity or quality). With regard to each irrigation technology it is assumed to be described by a fixed proportion production function. The costs are analysed first per unit of land and then corrected to per unit of output (the costs are 'indexed' to yield differences), thus relating the costs to the common denominator, a unit of the product. Therefore, the results are applicable to any scale of production.

Economic theory postulates that isoquants of efficient technologies are convex to the origin. The underlying reasons are the decreasing marginal rate of substitution between inputs and the daily observation that producing firms attain a stable internal equilibrium. Therefore, when estimating econometrically the parameters of a production function, one of the first tests that is performed on the estimated parameters is that of convexity of the implied isoquants. [3]

Before presenting the data the conclusions are stated. The data generate the expected convexity of the isoquants. Since the technologies are not continuous the isoquants are linear combinations of inputs for equal outputs (one specific is the unit output) of the efficient technologies. The data used, although they were taken from the Israeli experience, are within the ball park that represent prices prevailing in other countries. The only exception is the price of labour per day (or month). In developing countries it is considerably lower, but one should not forget to correct the per day cost to per output terms (to include productivity considerations, both of the worker and the entire system). However, even then the shape of the isoquant would not change, but would only rotate. [4]

The analysis below bypasses the issue of the price of water. However, the quantity of water and the water quality is explicitly embodied in the isoquant construction. Also, since the objective is that of cost minimisation, the price of output is ignored. This privilege does not hold when a technology is related to two or more crops systems, e.g., double-cropping. In this case one cannot bypass the issue of output prices and explicit assumptions have to be drawn. [5]

PREVIOUS WORKS IN THIS FIELD

The issue of efficiency in crop management is not new. It has occupied the agricultural economics literature not only in Israel but in various countries. Its presence usually reflects either a severe shortage of an input, e.g., water, or low profitability of crops due to low prices.

Sefarim and Shmueli (1975) examined the cost of water saved through the use of sprinkling and drip irrigation (defined as six distinct sub-technologies of that known more than a decade ago) in three field crops. Their main aim was to estimate the cost of saving water (substituting it by other inputs) and to compare it with the cost of water production (desalination or catching of floods). The isoquants they constructed for 'costs of irrigation' and 'amount of water' (see p. 1187) show explicitly the potential substitution between water and other irrigation costs and single out the inefficient technologies. The assumption hidden in their analysis is that all other costs of production do not differ between the different technologies.

Bergmann (1983; 1984) examined various systems in relation to each of the following cost factors: water consumption, investment costs, social factors, operation and maintenance costs, land use and ecological factors, as well as the yield per unit of land. It can be concluded, based on his evaluation of the relative ranking of the technologies (on a 1 to 7 scale), that it is impossible to state absolutely which irrigation system is the best. One reason is the nonquantifiable considerations of the decision-maker. Bergmann nevertheless emphasises that the internal rate of return of the investment is sensitive to all the above-mentioned factors, whereas the total investment costs is of relatively minor influence. Again, Bergmann regards profitability, i.e., the output price, as a crucial factor.

Caswell and Zilberman (1985; 1986) contributed significantly to the analysis of alternative irrigation systems. In their 1985 paper, they identify the factors that affect the share of land in returns (in fruit growing farms in California) for alternative irrigation technologies, mainly in relation to water pricing and water conservation. In the second paper (1986), they study the interrelationships between two farm characteristics, land quality and well depth, and the different irrigation technologies. Their findings explain why the various technologies were implemented in accordance with agricultural traditions and land quality.

A COMPUTATIONAL MODEL

The hereunder model is a formal framework for the computations which follow. One should first note that the minimal (or typical) land size on which different irrigation technologies are applied differ among technologies. A priori, one might also argue that the optimal land size is not necessarily the same for different techniques. Let us define by A_i this typical area (it may also be optimal, since it is determined by irrigation experts from competing companies). The two cost categories - *investment and labour* - are first calculated for these land sizes, and are denoted by I_i and L_i, respectively. Hence, the annual costs are the sum of the capital recovery costs, (K_i), and the costs of 'Labour' (L_i):

$$TC_i = K_i + L_i \tag{1}$$

Obviously, the capital recovery costs depend upon the investment, I_i, the interest rate, r, and years of life span that are relevant, T.

After normalising for the typical area covered by each system one obtains

$$AC(K) = K_i/A_i;\ AC(L) = L_i/A_i;\ AC = AC(K) + AC(L), \tag{2}$$

and the relative shares, of capital (CS_{i1}) and labour (CS_{i2}), respectively, for each irrigation technology, i, are

$$CS_{i1} = K_i/TC_i;\ CS_{i2} = L_i/TC_i. \tag{3}$$

They still require adjustments by *'correction indices'*: three correction reasons and thus indices were identified. A 'weighted index' was computed as their product. The values available for these indices are those of subjective judgements of experts; one may think of different ways to attain them, namely to come nearer to their exact values. They are,

A *yield index* (including water efficiency) Y_i *(4)*
(Y_i for sprinklers = 100),

An *expenditure effectiveness index* E_i *(5)*
(E_i for sprinklers = 100);

A *land utilization index* S_i *(6)*
(S_i for sprinklers = 100),

and *weighted index* $W_i = Y_i \cdot E_i \cdot S_i/10{,}000$ *(7)*
(obviously W_i for sprinklers = 100).

EMPIRICAL RESULTS

The isoquant analysis performed in this study examines four technologies that are already well established and a new one that is currently emerging.

Most of the data analysed here are based on the structure built by Joseph and Segal (1985), with the following modifications based on consultations with experts:

1. The investment in the fixed part of the drip system is much lower due to the latest innovations.
2. Repairs needed for moving lines were estimated annually as 6% of the investment, compared to 3% in their paper.
3. Minor corrections are related to labour inputs and to several investment components.

The already established technologies are: sprinklers, long-run drip, a linear-move irrigation system, and a center-pivot-move irrigation system. The new technology is the short-run drip technology. [6]

Table 1 contains: (1) a description of prior conditions for and basic features of each irrigation technology; (2) clarification of differences among the technologies, as in the similar and much more detailed table of Bergmann (1983, p. 50; 1984, p. 7).

On general grounds, the irrigation technologies are grouped into three equipment categories: sprinklers, moving lines and drip. It is clear that the five technologies differ from each other and the profiles described briefly in this table are complete vectors of characteristics. The sixth technology, for managing a 'controlled volume of roots', employs one of the drip irrigation techniques, and is conductive to a double-cropping system place of single crop (cotton) cultivation.

Table 1

Prior conditions for the utilisation and basic features of irrigation technologies

	Sprinklers	*Moving Lines*		*Drip irrigation*	
		Linear	*Pivot*	*Long-run**	*Short-run***
Prior Conditions					
Levelled area	-	+ (<10%)	+ (<10%)	-	-
Pressure of water supply	high	-	-	-	-
Wind	sensitive	-	-	-	-
Ease of Spraying Insecticides	+	+	-	-	+
Importance of the Water Quality	-	-	-	+	+
Skilled Labour Needs					
Current work	-	+	+	-	-
Setting and Rolling up	-	+	+	+	+
Control & Faults	high	med.	med.	low	low
Utilisation of Marginal & Irregular Areas	med.	low	poor	high	high

* Existing tubes of rigid walls.
** Thin walls, under or above the surface.

Table 2 presents the cost components of each technology and their conversion to a standard input unit, 100 hectares of land. The costs are calculated for Israeli conditions and data.

Part A of this table portrays the cost breakdown of existing irrigation systems that are sold in Israel. Its labour component is based on local costs of skilled and unskilled labour (6.5 and 2 US$ per working hour, respectively). The distribution of their employment was estimated by the authors. The investments include the initial cost of the equipment plus the cost of repairs. The latter costs are estimated as 6% of the former costs for the moving lines and 3% for the other techniques. These capital investments should be recovered during the life span of the technique. The life span is the weighted average of the life spans of all equipment components. Since in the 'short-run' drip, investment components such as fertiliser tanks and injectors, filters and wide, rigid supply lines have a major share (54%), and they depreciate over 15 years, the weighted average life span of the entire technique amounts to 8 - 10 years. The minimal figure was chosen for calculations in order to reduce risk.

Table 2

Costs components of irrigation technologies

	Sprinklers	*Moving Lines*		*Drip Irrigation*	
		Linear	*Pivot*	*Long-run*	*Short-run*
A. Basic Data					
Life span of investment (years)	8	10	10	11	8
Investment, incl. repairs (US$ 1000)	36.4	82.6	63.6	56.2	24.2
Irrigated area (ha)	36	69	36	36	36
Labour and remaining costs (US$ 1000)	5.9	2.4	1.0	4.5	4.6
B. Annual Costs Calculated for a Standard Unit of 100 Hectares (one season of use in cotton growing)					
Capital invested (US$ 1000)	101.1	119.8	176.7	156.1	67.2
Capital recovery at 8% (US$ 1000)	17.6	17.8	26.3	21.9	11.7
Labour and the rest (US$ 1000)	16.4	3.5	2.8	7.0	7.5
Total annual costs (US$ 1000)	34.0	21.3	29.1	28.9	19.2
Share of Capital Recovery of the Annual Cost (%)	52	84	90	76	61

The main outcome of this presentation is the capital intensity of the various technologies: the least intensive is the traditional irrigation through sprinklers. This finding is at variance with the established assumption that drip is highly capital intensive. The 'short-run' drip versions, that are expected to be the future technology, are close to sprinklers in terms of capital intensity.

Table 3 shows the sensitivity of the costs to the assumed interest rates, to the life span of the equipment and to what is defined as the 'development option' - a combination of longer recovery period and lower interest rate (2%).

It is evident that all the alternatives are preferred (less costly than) over sprinkling irrigation. Thus one may ask, 'why don't farmers immediately change their irrigation technology'?

The answer probably lies in the share of the fixed costs component: in the existing nondeteriorated equipment, in the ruling traditions, in beliefs and in institutions. All these explain the 'rationality' of the observed behaviour of the farmers.

Table 3

Sensitivity analysis of average annual costs of the different irrigation technologies (US$/ha)

	Sprinklers	*Moving Lines*		*Drip Irrigation*	
		Linear	*Pivot*	*Long-run*	*Short-run*
A. *Capital Recovery at Different Interest Rates:*					
Annual (8%)	340	213	291	289	192
Index (sprinklers=100)	100	63	86	85	56
Annual cost (2%)	302	168	224	229	166
Index (sprinklers=100)	100	56	74	76	55
Annual cost (10%)	382	265	367	356	220
Index (sprinklers=100)	100	69	96	93	58
B. *Life Span of Investment*					
Change (from - to), years	8-12	10-14	10-14	11-17	8-11
Annual cost (8%)	298	180	242	241	169
Index (sprinklers=100)	100	61	81	81	57
C. *'Development Effort'* (longer life and subsidised capital - 2%):					
Annual cost	260	134	174	179	143
Index (sprinklers=100)	100	52	67	69	55

It can be seen that the changes in costs that originated from the modification of the interest rate are positively correlated with capital intensity and with the expected life span of the equipment. In view of these results it should be remembered that the innovations in this field are orientated to lengthen the systems' life span.

In Table 4, efficiency and productivity indices are introduced; they are not embodied in the cost calculations made above but affect the per unit of output costs.

These indices are based upon views of experts and on results of experimental work that was carried out in Israel over a long period. Nevertheless, some clarifications are necessary: the pivot system has a yield index higher than the ratio between the geometric forms of a circle and a quadrangle; it is an outcome of the efforts to irrigate most of the area between the 'circles' or of better water efficiency. The drip irrigation system has an advantage in the utilisation of marginal areas of land. It is especially valuable for small areas and in hilly regions, whereas its importance is much smaller on a plateau. These estimates are supported by the results of a similar research performed in the USA by Wilson *et al.* (1984). The base for all indices is sprinkling. The 'weighted index' is the multiplication of the three efficiency factors: yield, expenditure effectiveness and land utilisation.

Table 4

Productivity and efficiency indices of the various irrigation technologies

	Sprinklers	*Moving Lines*		*Drip irrigation*	
		Linear	*Pivot*	*Long-run*	*Short-run*
A. Yield index	100	90-95	100-105	115-120	125
B. Expenditures' effectiveness index	100	95	95	105	100
C. Land utilization index	100	100	90	110	110
D. Weighted index (D = A•B•C)	100	86-90	86-90	133-129	151

Table 5 presents the costs per unit of output adjusted for efficiency, employing the indices presented in Table 4.

Table 5

Annual costs per hectare, corrected by indices (US$/ha)

	Sprinklers	*Moving Lines*		*Drip Irrigation*	
		Linear	*Pivot*	*Long-run*	*Short-run*
A.1 Regular interest rate (8%) and life span - yield index:					
- Capital	176	192	257	186	94
- Labour, etc.	164	38	27	107	102
Total costs	340	230	284	293	196
A.2 Regular interest rate (8%) and life span weighted index:					
- Capital	176	202	299	161	77
- Labour, etc.	164	40	32	92	85
Total costs	340	242	331	253	162
B.1 Subsidised capital (2%) and prolonged life-yield index:					
- Capital	96	107	142	93	55
- Labour, etc.	164	38	27	107	102
Total costs	260	145	169	200	157
B.2 Subsidised capital (2%) and prolonged life-weighted index:					
- Capital	96	112	166	80	45
- Labour, etc.	164	40	32	92	85
Total costs	260	152	198	172	130

One can see the increased spread of the costs over technologies when adjusting the costs by the weighted efficiency index.

The split of the costs between 'labour' and capital is used below for construction of the production isoquants. Figure 1a exhibits the isoquants in two versions which differ in the correction index; one is indexed only by a yield index while the other is adjusted by the weighted index of efficiency. Both are for a regular life span, and an interest rate of 8%. Figure 1b expresses the same ideas as Figure 1a, but under 'development conditions', i.e., offering the investors the needed credit at an interest rate of only 2% and over a longer recovery period.

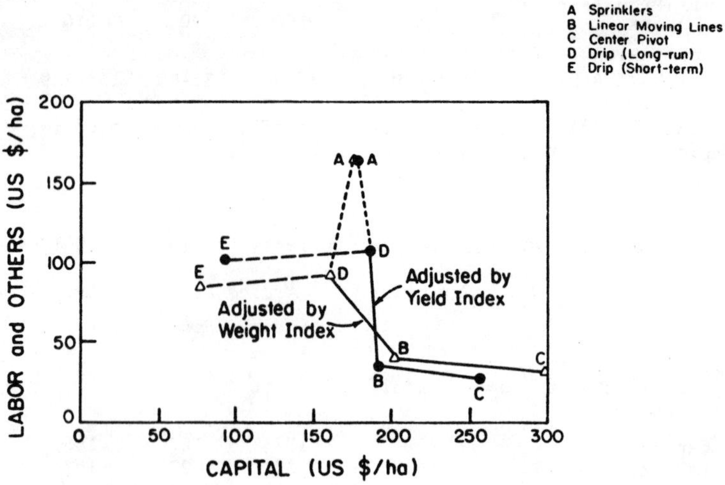

Figure 1a Irrigation technologies isoquants: regular-term technology (8% interest rate)

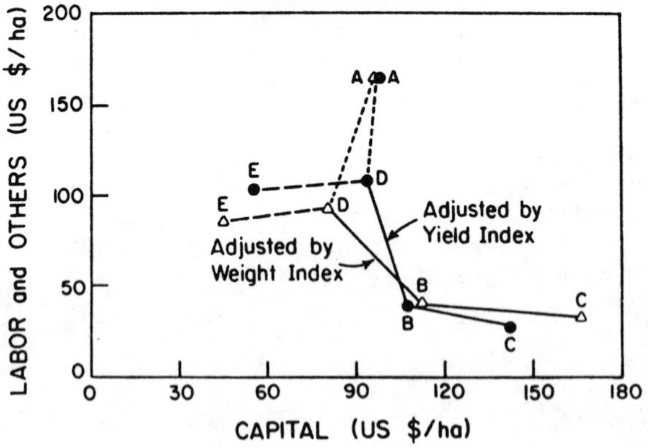

Figure 1b Irrigation technologies isoquants: long-term technology (2% interest rate)

There are no significant differences between the two isoquants presented in Figures 1a and 1b.

CONCLUSIONS AND AN EPILOGUE: THE SIXTH TECHNOLOGY

Analysis of the isoquants for growing cotton reveals that moving lines and long-run drip are on the same isoquant. Sprinklers are redundant at any combination of input prices, i.e., it should be phased out, at least with respect to cotton growing under Israeli conditions, after the existing equipment is fully exploited. Short-run drip is seen as the best method in terms of costs, thus it is likely to be the future irrigation technology - again for cotton and in Israel. These two reservations are made because, as of now, there are only partial confirmations for other agro-economic conditions. It is interesting to note, that these findings are already observed in the behaviour of farmers: the distribution of cotton areas over the last years between those irrigated by drip and those by sprinklers has changed rapidly to drip (Fishelson and Rymon, 1986).

The sixth technology is a combination of one of the drip versions with better agronomic management and tighter irrigation control; this is the approach of 'controlled volume of roots' presented by Carmi and Plaut (1986): double-cropping under temperate conditions is achieved by limiting the volume of the roots' system and inducing the pass to the reproductive stage of the cotton plants. The result is a shortening of the period the cotton plants stand in the field, till it releases the field for another crop, say winter wheat.

The economics of double-cropping when the two crops are different deserves a short specific discussion. It is obvious that in the double-cropping case one has to consider the prices explicitly, while in the case of a single crop only the yield per unit land is of concern. Since the two crops are traded on international markets, the prior information obtained on expected price helps to smooth the fluctuations in total income via the shifting from one crop to another. Cotton still remains - at least under Israeli conditions - the main crop. By adding a second crop, e.g. wheat, the cotton output is lowered. However, experiments have shown that the grain crop contributes an additional revenue of US$ 500 to 1,000 per hectare, while the additional costs plus the value of the loss of cotton yield is estimated to be below US$ 500.

This 'sixth technology' is a new approach first demonstrated in Israel for cotton and wheat, but not limited to these crops. It is believed, that in the future the concept of combining the highest technical and professional capacities to control the plant development will gain its proper position as a way of maximising achievements from the biological system.

A comprehensive analysis of such a basket of options - portrayed here under specific existing Israeli constraints - will maximise the income farmers will generate.

NOTES

* We would like to thank our student. Mr. Paz Shanni for data collection. We acknowledge valuable comments from Mr. Uri Or of Kibbutz Magal, Dr. Ariel Dinar of the Faculty of Agriculture of the Hebrew University of Jerusalem and Dr. Hellmuth Bergmann of the European Investment Bank (Luxemburg). Financial assistance was partly provided by the Foerder Institute of Economic Research, Tel Aviv University, Israel.

** Department of R & D Economics, Agricultural Research Organization, The Volcani Center, Bet Dagan, Israel.

*** Department of Economics, Tel Aviv University, Ramat Aviv, Israel.

[1] Wilson *et al.*, (1984), attached a caution (p. 6). Because drip irrigation of cotton is a new technology, some of the information available about drip is tentative.

[2] In other countries where irrigated farming is the dominating practice, the share of cotton is considerably low. Yet, there are particular areas where it becomes the main crop, e.g., Southern California, Arizona, Texas, South Brazil and similar agro-environments.

[3] For some production functions this is not necessary, since the signs of the estimated parameters already tell the story (e.g., Cobb-Douglas, CES). In others, e.g., Translog, an explicit check for local convexity is performed (e.g., Berndt-Wood (1975)).

[4] When dealing with the issue of labour cost in developing countries, a common mistake is to forget when and where a skilled worker is needed; such an input might be scarce and costly.

[5] The only time that one can still bypass the price of output is when the prices of the various products in the system are strongly interrelated (move proportionally in the same direction). This is obviously very unlikely to happen, if consideration is not bound to a group of crops of common market features.

[6] This new technology has already been tested in the past, but the version was not good enough (technically) to be adopted on a large scale. Technical improvements that are currently being introduced will necessitate re-analyse of its potential.

REFERENCES

Bergmann, H., (1983), 'Economic considerations in choosing irrigation systems', *Proceedings of the 3rd International Conference on Irrigation*, Tel Aviv, Israel, 47 - 57.

Bergmann, H., (1984), 'Economic considerations in selecting irrigation systems', *Water and Irrigation Review*, 5 - 9.

Berndt, E.R. and Wood, D.O., (1975), 'Technology, prices, and the derived demand for energy', *Rev. Econ. Stat.*, 57: 259 - 268.

Carmi, A. and Plaut, Z., (1986), 'A new concept of a double-cropping system, cotton-wheat based on frequent drip irrigation and control of the root zone volume', Lecture at the *1st International Symposium on Optimal Yield Management*, Tel Aviv, Israel.

Caswell, M. and Zilberman, D., 'The choices of irrigation technologies in California', *Amer. Jour. Agr. Econ.*, 67(2): 224 - 234.

Caswell, M. and Zilberman, D., (1986), 'The effects of well depth and land quality on the choice of irrigation technology', *Amer. Jour. Agr. Econ.*, 68(4): 798 - 811.

Fishelson, G. and Rymon, D., (1986), 'Adoption of agricultural innovations, the case of drip irrigation of cotton in Israel', *Working Paper No. 35-86, The Foerder Institute for Economic Research,* Tel Aviv University, Tel Aviv, Israel.

Joseph, S. and Segal, D., (1985), 'Economic analysis of irrigation systems', *Agricultural Research Organization, The Agricultural Engineering Institute,* 31 pp. (in Hebrew).

Sepharim, Y. and Shmueli, M., (1975), 'The cost of water saved by using alternative irrigation systems', *Hassadeh,* 58(7): 1179 - 1194 (in Hebrew).

Wilson, P., Ayer, H. and Snider, G., (1984), 'Drip irrigation for cotton', *Agr. Econ. Report No. 517, USDA.*

21 Economic considerations as affected by quantity–quality price system incentives for processing tomatoes

ARIEL DINAR AND DAN MAROM

INTRODUCTION

The increase in the standard of living of the developed society has caused changes in consumer tastes. This of course has led to market preferences reflected in higher prices for quality or for products preferred by consumers which are produced from the same raw material. Examples can be given for almost every agricultural crop.

The pricing of quality attributes is the domain of the hedonic price theory (Griliches, 1971) which analyses the valuation of goods and services based on their specific characteristics. Most hedonic price studies have dealt with the pricing problem from the consumers' point of view (Griliches, 1971). Less emphasis has been given to establishing efficient incentives to the producer to provide the characteristics (quality) requested by the market (Golan and Shalit, 1985; Gorman, 1980).

High quality of an agricultural product is usually associated with higher production costs, especially when the final product has to be processed after harvest. Hence, an adequate incentive for quality is one of several measures encouraging the individual producer to be quality oriented.

The food industry often differs from other manufacturing firms. Food processing is generally highly seasonal, the raw product may be perishable in its fresh state and input quality may be variable (Logan, 1984). Tomato processing follows the trend outlined in general terms above. Tomatoes can be converted into various products, such as whole peeled tomatoes, sauce, paste, puree, tomato juice and ketchup. Each product has its quality requirements from the raw material and the related production costs.

Each tomato manufacturer (processor) has a given products mix derived from market preferences which determine the consequential desired quality requirements of raw material. The quality requirements can be met by tomato growers who supply the raw material for the tomato industry. This can be done by using an appropriate quality incentive price system which will be attractive for both growers and processors.

The aim of this chapter is to describe the economic aspects related to processors and growers and to analyse factors involved in the decision process made by the individual tomato grower under conditions prevailing in Israel. The paper is divided into two parts: the first part describes the industry (processors and growers) and the pricing and quality systems; the second part analyses several economic aspects derived from the quality price system incentive. The introduction is followed by a short review of the development of the tomato processing industry in Israel. Next, the characteristics and interests of processors and growers are typified and discussed. Following these sections is an economic model providing the conditions for optimal production from the growers' point of view. The model is applied to conditions prevailing in one of the important growing tomato regions located in northern coastal Israel. Results are presented and discussed.

PART ONE: PROCESSING TOMATOES IN ISRAEL

About 250,000 tons of tomatoes were supplied to the processing industry in 1985. During recent years production has declined mainly due to the increase in labour cost and profitability of competing crops, on the one hand, and keen competition in the world market of tomato products, on the other hand.

Israel has a very small share (1.6%) in the world production. Most of the products (85%) are exported, mainly to the USA (62%). The value of exported products in 1985 was 29 million dollars (Israel Vegetables Production and Marketing Board (IVPMB), 1985). In 1983, Israeli processed tomatoes constituted 18.8% of the imports to the USA (Ben-David, 1985). Processing tomatoes occupied 3,500 ha in 1985 (3% of the irrigated field crops in Israel).

The annual capacity of the 16 factories in the tomato processing industry is 500,000 ton, but this capacity is only partially utilised. Production is concentrated; four factories produce 70% of the products. Factories process tomato during the summer and citrus during the winter.

The processing industry

The tomato processors are organised in the Israel Tomatoes Products Board (ITPB). Most of the processing plants are owned by farmers' organisations. Since 1983, when highest volume of production, 290,000 ton, was processed (IVPMB, 1985), there has been a declining trend in produced volumes, which reached 250,000 tons in 1985.

There is no clear cut specialisation among the processing plants. All factories produce the same product categories though in different proportions. The average country product composition is presented in Table 1.

Table 1

Composition of production by product categories for 1985
(percentage by value of production)

Product category	*% of production*
Crushed and whole peeled	34.5
Sauce	32.8
Puree	23.3
Juice and ketchup	9.4
	100.0

Source: IVPMB, 1985.

As mentioned above, factories usually use the same equipment to process tomatoes in the summer and citrus fruits during the winter. The prolonged operation period provides the Israeli processing industry a cost advantage over factories specialising in tomato only.

The Israeli industry is responsive to market opportunities. Because of the small sizes of the factories and their relatively wide range of products, they make quick decisions on the proportions of products in their product mix. Usually the production is for direct marketing according to already available orders. In order to improve its competitiveness, the industry also strives to achieve a higher standard for product quality (Garoyan, 1984). The dynamic nature of the tomato industry is demonstrated in Table 2, which shows that over a relatively short period the product composition has been changed and export rates have been increased; however, keen competition on the part of Italy, Spain, Portugal, Taiwan and Mexico in the American market might cause a decline of its importance for the Israeli tomato industry.

Table 2

Changes in product mix and exports to USA during 1981 - 85

Product category	*% of export*	
	1981	1985
Crushed and whole peeled	29.9	34.4
Sauce	23.9	32.8
Puree	18.2	23.3
Juice and ketchup	27.8	9.4
	100.0	100.0
Share of exports to USA	50.8	61.5

Detailed information regarding cost structure of the processing industry is not available. The price of $285/ton was cited in 1985 as a basis for tomatoes price negotiation. This is the factory cost for processing a standard product containing 30 Brix (Israel Ministry of Agriculture, 1985). The component of raw material in the product cost varies according to the ratio of raw material needed for a given final product. Average values are presented in Table 3.

Table 3

Raw material (ton) needed for one ton of product

Product category	*Raw material needed for one ton*
Crushed and whole peeled	1.2
Sauce	3.3
Puree	7.0
Juice and ketchup	1.4

The energy cost component is directly related to the raw-final product ratio since energy is mainly used for dehydration of fresh tomatoes. For production of 30 Brix (ratio 1:7), energy constitutes 10 - 15% of factory cost (excluding cost of purchasing raw material).

The farm producing sector

Processing tomatoes are mostly (67%) grown by individual farmers organised in cooperatives called moshavim, 11% are produced in collective farms (kibbutzim), and the rest by private farmers. Some 85% of the area is drip irrigated enabling water and liquid fertilisers to be applied precisely in suitable amounts and at appropriate timing. The crop is planted or sown from February to April, depending on the region (earlier in hot regions). Harvesting is concentrated during seven weeks in July and August when 80% of the product is picked.

The average yield in 1983 was 76 ton/ha (Institute of Farm Income Research (IFIR), 1984). Only 10% of the area is mechanically harvested; this is mainly due to the high prices of imported machinery and the relatively small size of tomato plots in the moshav farms (1 - 5 ha), which makes the use of tomato picking combines unprofitable.

The country average of gross margin (returns to land, management and capital) reported in 1983 was $ 950/ha and the reported break-even tomato yield (the yield needed to cover production costs) was 75 ton/ha (IFIR, 1984).

Tomato growers are organised as a special section within the Israel Vegetable Growers Organisation (IVGO). The IVGO negotiates prices with the organisation of the processing industry.

Annual contracts between individual factories and growers are the basis for production planning for both sides.

Israeli farmers, including tomato growers, are well known for their quick response to technological innovations. Farmers either individually or through their organisations are not only financing applied research but also conducting experiments on their plots. The processing industry also has an important role in planning and financing agricultural research. The Israel Extension Service transfers new knowledge through seasonal recommendations to be made available to the growers (Geizenberg and Etiel, 1987).

Quality factors for processing tomatoes

The most important quality factors of the tomato fruit used for processing are: total soluble solids (TSS), acidity, colour, consistency of the juice, shape and structure of the fruit. For different tomato products, different factors are the most important ones (e.g., for concentrated products TSS is the main factor, for whole peeled tomatoes the colour and fruit shape are the important factors). TSS had been found to be the most single significant factor which determines the quality of the final product.

A higher value of TSS in the fruit does not only contribute to the flavour and quality of the final product, but also significantly reduces the processing production costs. This is due to the lower energy costs needed for dehydration and also to the fact that a lower tonnage of raw material is needed to produce a particular final product.

The pricing system

Prices are negotiated every year, before the season, for all growers and processors represented by their organisations. Prices are specified in US dollars, and premiums are established for early and late crops.

Since 1985 a premium has also been paid according to TSS content (IVPMB, 1985) specified in Brix units, and hereafter is referred to as the quality incentive price system (Table 4). The planning authority of the Ministry of Agriculture has played a major role in initiating the quality price system incentive. The rationale for its improvement was to encourage economic use of water and other scarce production factors in Israeli agriculture.

Table 4

Quality incentives for processing tomato

Brix level (%)	*Fine ($/ton)*	*Brix level (%)*	*Bonus ($/ton)*
4.5 - 4.6	-2.50	4.7 - 4.8	0.00
4.3 - 4.4	-5.00	4.9 - 5.0	3.25
4.1 - 4.2	-7.50	5.1 - 5.2	6.50
3.9 - 4.0	-10.00	5.3 - 5.4	9.75
3.7 - 3.8	-12.50	5.5 - 5.6	13.00
3.5 - 3.6	-15.00	5.7 - 5.8	16.25
3.3 - 3.4	-17.50	5.9 - 6.0	19.50
3.1 - 3.2	-20.00	6.1 - 6.2	22.75
less than 3.0	-22.50	6.3 and over	26.00

Values are September 1985, constant dollars.

Source: IVPMB, 1985.

From the growers' point of view, prices reflect cost of production and profitability of alternative crops. From the processors' point of view it reflects the potential market value for the planned product mix.

In Israel there is only one quality incentive price system for the entire industry, based on TSS content only. In the USA, each factory determines its own quality incentive program (California Tomato Growers Association, 1986) dependent on its product mix; each quality incentive system includes TSS as one element but most often it is accompanied by one or both of the following criteria: percentage of limited use and tomato variety group.

PART TWO: ECONOMIC CONSIDERATIONS OF THE INDIVIDUAL GROWER

The first part of this paper has provided some insight into the Israeli tomato industry. This is a complicated and sophisticated system which is exported and quality oriented. Many decisions are made in a sensitive production process which links together processors and growers, sometimes with conflicting interests. An annual planning model for the industry (Logan, 1984) might regulate the production and improve the industry's efficiency. However, here the analysis is directed towards the field level only. The model is for an individual grower who usually faces given conditions (e.g., input-output prices) and his decisions involve the allocation of inputs in the field in order to improve his own welfare.

Model

Consider a plot of 1 ha with limited available water and other inputs used for growing processing tomato. The grower is assumed to be a profit maximiser. Assuming a full knowledge of the parameters involved in the production process and the given input-output prices, the grower has to make decisions regarding: the season (early or late), the irrigation system (drip or sprinkler), the tomato variety (M-82 or another), and the seasonal applied irrigation water. The grower is aware of these means by which he can control his production and improve profitability.

For simplicity assume that the grower has already chosen the optimal season, irrigation system and tomato variety. The remaining decision variable affecting production level is the seasonal applied irrigation water. The grower controls production mainly by changing water quantities during certain phenological growth stages of the tomato plant (Rudich *et al.*, 1985; Luchinsky, 1986).

Consider first the case where only a single basic product price for quantity exists. The grower's profit is

$$\Pi_1 = P^y Y(W) - C(W).$$

Here P^y is the price for tomato yield ($/ton) net of marketing cost, $Y(W)$ is the tomato water production function in units of ton/ha. $Y(W)$ is monotonic with $\partial Y/\partial W > 0$ and $\partial^2 Y/\partial W^2 < 0$. $C(W)$ is the variable cost function with $\partial C/\partial W > 0$ and $\partial^2 C/\partial W^2 < 0$.

First order conditions for optimality with respect to W provide

$$P^y \cdot \partial Y/\partial W = \partial C/\partial W. \quad (2)$$

The optimal solution W_1^* is such that the value of marginal productivity equals the marginal cost. Consider now the case where a quality incentive price system $[P^b(B)]$(\$/ton) exists, where B is a measure of quality, along with the regular basic price for quantity.

The grower's profit is now

$$\Pi_2 = P^y Y(W) + P^b(B)Y(W) - C(W) \quad (3)$$

and

$$B = f[Y(W)]. \quad (4)$$

Here $P^b(B)$ is the incentive price system. $P^b(B) \lesseqgtr 0$, the function f is a monotonic decreasing function with $\partial B/\partial Y < 0$ and $\partial^2 B/\partial Y^2 < 0$.

First order conditions for optimality with respect to W provide (after arrangement)

$$\partial Y/\partial W[Y(W) \cdot \partial P^b/\partial B \cdot \partial B/\partial Y + P^b(B)] = \partial C/\partial W \quad (5)$$

where $\partial P^b/\partial B$ is the incremental income from an additional increase in quality (> 0); $\partial B/\partial Y$ is the technical rate of substitution between yield and quality (< 0).

The expression in brackets (equation *(5)*) must be positive in order to obtain an optimal solution. Since $\partial B/\partial Y < 0$ this condition means that

$$P^y + P^b(B) > Y(W) \cdot \partial P^b/\partial B \cdot \partial B/\partial Y$$

and after rearrangement becomes

$$\frac{P^y + P^b(B)}{\partial P^b/\partial B} > \frac{Y(W)}{\partial Y/\partial B} \quad (6)$$

The left-hand side of equation *(6)* is the ratio between the total income per ton of tomato of a given quality and the marginal income from quality changes. The right-hand side is the ratio between the yield and the reciprocal derivative of the rate of technical substitution between yield and quality. In order to fulfill equation *(6)* and since $P^b(B) \lesseqgtr 0$ one requests that $P^y > P^b(B)$ if $P^b(B) < 0$.

Having in mind the previously mentioned characteristics of Y, B, P^b and $C(W)$ it can be easily shown that the optimal solution W_2^* to equations *(3)* and *(4)* which fulfills equations *(5)* and *(6)* is obtained such that $W_2^* \leq W_1^*$ for the same environmental conditions. Increasing P^y while holding all other variables constant results in an increase in the optimal applied water. Increasing the slope of the variable cost function (e.g., increasing water price) while holding all other variables constant results in a decrease in the optimal applied water. An increase in the slope of the quality incentive price system while holding all other variable constant results in a decrease in optimal water applications.

The model can be applied to evaluate optimal decisions of a grower under different conditions and the consequential expenditures of the tomato industry for buying raw material.

The model is applied with some empirical specifications to conditions prevailing in an important growing tomatoes region in Israel.

Data and empirical specifications

The empirical analysis is applied to the Western Galilee region. Processing tomatoes are widely grown in this region which is located at the north-western coastal area of Israel. The region is characterised by a rainy winter (600 mm/yr) and a hot growing season (April to August) with pan evaporation values around 700 mm.

The region has set unique interrelationships among growers, extension specialists and researchers. An experimental station sponsored by the growers is located in the region and carries out field experiments aimed at improving knowledge and solving problems in the agricultural production process.

Field experiments carried out during 1980 - 1984 in this experimental station serve as a basis for establishing the quantitative relationships between applied water and yield quantity and quality (Rudich *et al.*, 1985). Other input values and costs of production of tomatoes in the region are based on crop budgets provided by Agami and Eshet (1985). Yield prices and quality incentives were provided by the tomato industry (IVPMB, 1985). All prices are September 1985 constant dollars.

Using the same vocabulary as in the model section, the empirical profit function to be maximised is

$$\Pi = (1 - \alpha)[P^y Y(W) + P^b(B)Y(W)] - P^h Y(W) - P^w W - C - L \qquad (7)$$

where $B = f[Y(W)]$.

The profit Π is calculated as returns to land, management and capital for irrigation equipment ($/ha); α is a factor taking into account dues and taxes as a percentage of revenue ($\alpha = 3.1\%$); P^y is the basic price paid by the industry for one ton of tomatoes ($ 63); for simplicity assume that payments are made on harvest day. The set of incentive prices $P^b(B)$ ($/ton) for quality [in terms of Brix (B)] is presented in Table 4. Harvesting and transporting price (P^h) is $ 21/ton, and water price (P^w) is $ .11/m^3. Production costs not directly related to water (C) include seed, fertilisers, machinary etc., and are assumed to be constant under all situations considered. The value assigned to these costs is $ 1,596.4/ha. Labour cost (L) excluding harvesting is calculated assuming 15 skilled and 20 casual working days per ha with corresponding rates of $ 35/day and $ 21/day. This has resulted with labour cost of $ 945/ha.

A linear yield (quantity) - water production function for the M - 82 variety under drip irrigation is used based on Rudich *et al.* (1985).

$$Y(W) = 30.4 + .0214W.$$

Here W is the seasonal applied water (m^3/ha). The level of total soluble solids in the yield is measured in terms of Brix (B). (B is in percent; in order to convert it to TSS units per ha one has to multiply B by Y). A linear relationship is used based also on Rudich *et al.* (1985)

$B(Y) = 10.428 - .0635Y$

The empirical model is applied to the region under investigation and provides optimal values for water application (W^*), yield (Y^*), Brix level (B^*) and profit (Π^*). For several purposes described later a variable measuring total soluble solids (TSS) per ha is also calculated

$$\text{TSS} = B \cdot Y.$$

Industry payment (IP) for the raw material is also calculated separately

$$\text{IP} = [P^y + P^b(B)] \cdot Y.$$

The model first analyses the current situation (base run) using the values mentioned above. Then, several sensitivity analyses are made aimed at evaluating assumptions, situations and the sensitivity of the results to additional scientific information. The sensitivity runs include (holding all other variables constant) :

1. Changes in water price P^w;
2. Changes in the basic tomato price P^y;
3. Change in the slope of the yield - water production function representing an introduction of new variety or a better irrigation practice;
4. Change in the slope of the substitution curve between Brix and yield;
5. A situation without quality incentives, $P^b(B) = 0$;
6. Introduction of new scientific information based on additional field experiment results (non-linear estimates for Y and B).

In addition to the sensitivity runs, two more analyses are performed. In the first the marginal cost (MC) for producing TSS is calculated using the current data as a demonstration for a possible different pricing system. The second analysis is aimed at demonstrating the potential embodied in the quality incentive price system. Alternative values for the incentive system (P'^b) are suggested such that grower's profit increases, industry payment to the grower are not increased and water consumption is decreased. The comparison will include a case where $P^b(B) = 0$ and P^y has the current value ($ 63/ton) against a case where P^b and P^y have the current value (base case).

Results

The computed values for optimal applied water (W^*) and associated profit (Π^*), yield (Y^*) and Brix (B^*) are presented in Tables 5 through 7 for the analysed conditions.

Table 5

Evaluation of the 'Rudich *et al.*' - based production functions under different scenarios and parameter values

	W^* (m^3/ha)	Π^* (\$/ha)	Y^* (ton/ha)	B^* (Brix)
Base case	1,800	1,475	68.9	6.05
P^W = \$ .50/m	1,600	784	64.6	6.32
P^Y = \$ 90/ton	2,100	3,340	75.3	5.64
$\partial Y/\partial W$ = .0150	2,400	1,457	66.4	6.21
$\partial Y/\partial W$ = .0250	1,400	1,506	65.4	6.27
$\partial B/\partial Y$ =-.0550	2,300	2,089	79.6	6.04
$\partial B/\partial Y$ =-.0700	1,400	1,174	60.3	6.20

The results for the base case (Table 5) suggest an optimal water application of 1,800 m^3/ha which is associated with a value of \$ 1,475/ ha for returns to land, capital and management. This solution is derived from the shape of the production functions and the input-output prices. The optimal solution is not sensitive to changes in water price in the range from \$ .11/m^3 to \$ .49/m^3 but when the water price hits \$ .50/m^3 the optimal water application is reduced and so are the returns for land and management. Additional runs made for higher values of water prices have resulted with lower water application (not presented).

The effect of increasing the P^Y value on the optimal decision was also analysed. The optimal solution is not sensitive to changes in P^Y in the range from \$ 63/ton to \$ 89/ton. A price of \$ 90/ton has caused an increase in the optimal water application from 1,800 m^3/ha (in the base case) to 2,100 m^3/ha.

The production functions used in this analysis are based on field experiments providing data for a statistical estimation. Although the estimations (Rudich *et al.*, 1985) are statistically significant, some precaution measures should be taken before using them for practical production decisions. Such measures are the sensitivity analyses for the estimated values of the slopes of the production functions presented in Table 5. The results indicate that the optimal solution is very sensitive to the slope of both functions [$Y(W)$, $B(Y)$]. A change of about 25% in its value to both sides is accompanied by a similar change in the optimal value of water application.

Luchinsky (1986) provides experimental data for estimating yield-water and yield-quality relationships for the drip-irrigated M-82 variety. The following regression estimates (based on Luchinsky (1986)) were used in the analyses in addition to those based on Rudich *et al.* (1985) :

$$Y = 52.17\ W^{.043} \qquad R^2 = .710$$
$$B = 82.51\ Y^{-.602} \qquad R^2 = .989\ .$$

All estimated coefficients are significant ($P < .05$).

Table 6

Evaluation of the 'Luchinsky'-based production functions under different scenarios

	W^* (m^3/ha)	Π^* (\$/ha)	Y^* (ton/ha)	B^* (Brix)
Base case	1,800	1,895	72.0	6.28
$P^w = .20$	1,000	1,775	70.2	6.38
$P^w = .30$	600	1,695	68.7	6.46
$P^y = 70$	2,000	2,385	72.3	6.26
$P^y = 80$	2,300	3,087	72.8	6.24

The production functions based on Luchinsky are also analysed under several sensitivity analyses. The results are presented in Table 6. Scrutiny of Table 6 indicates that the non-linear production functions are much more sensitive to changes in water and product prices.

Table 7

Welfare comparisons of situations with alternative quality incentive price systems

	Y^* (ton/ha)	B^* (Brix)	W^* (m^3/ha)	Π^* (\$/ha)	*IP* (\$/ha)	*TSS* (ton/ha)
$P^b(B)=0$; $P^y = 63$	105.3	3.74	3,500	1,242	6,634	3.94
[1]base case	68.9	6.05	1,800	1,475	5,910	4.17
[1]$P^b(B)=P'^b$; $P^y = 63$	64.6	6.32	1,600	1,585	5,910	4.08

[1] The quality incentive price system is presented in Figure 1 for the base case and for the alternative system P^b.

The results for the case where $P^b(B) = 0$ are presented in Table 7. In the absence of a quality incentive price system, the optimal water application is far above the base case solution. The production of TSS (for which industry pays) is lower than in the base case, grower's profit is smaller and the industry payments (IP) to the grower for purchasing the raw material are much higher.

The results of applying an alternative quality incentive price system (presented in Figure 1) are presented in Table 7. This system (P'^b) differs from the current one only in the incentives rates (fines are the same as in the current one). The optimal solution derived from applying such a system is characterised with almost the same TSS yield per ha (differences are due to computational rigidities of the computer program caused by the water application interval), the same industry payment and higher profits to the grower. The grower also benefits from water saving as compared to the base case.

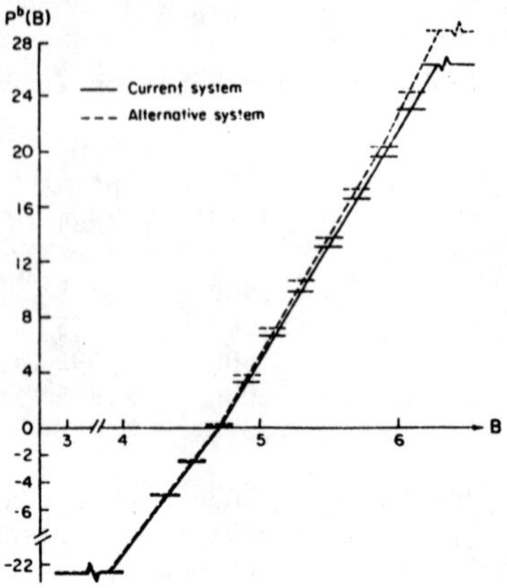

Figure 1 Description of two possible incentive price systems for processing tomatoes

Since the tomato yield is translated into TSS units (Figure 2) the results obtained by the marginal cost (MC) function of producing TSS yield can provide the TSS supply function (Figure 2).

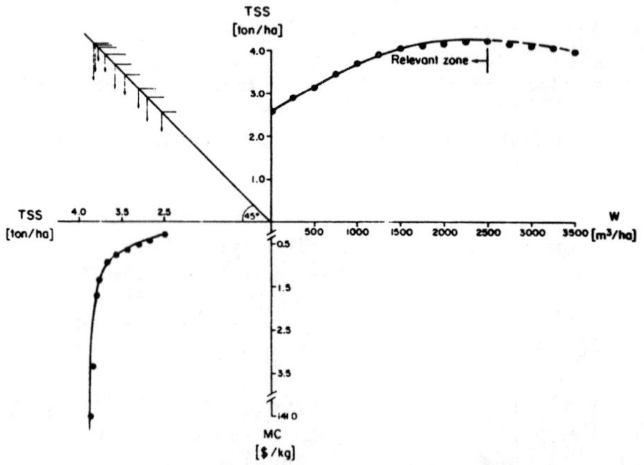

Figure 2 The supply function of total soluble solids for one ha of processing

Notice that the slope of the supply curve is moderate up to a production level of TSS = 4 ton/ha but is very steep in the range from 4 ton/ha to 4.5 ton/ha.

The system presented in Figure 2 can be considered as an alternative price system for the processing tomato growers. Since the yield measured by TSS units is the input requested by the industry, it makes sense that the price which the grower faces would be for TSS yield.

Therefore, given the supply curve of TSS and the price the processor offers the grower for 1 kg of TSS, the grower can derive his production level and the input level needed to achieve that TSS yield.

For example (see Figure 2) suppose that the price is $ 1.4/kg of TSS. Then by projecting from the price axis to the TSS axis (the MC curve), the grower finds the level of production to be 3.75 ton of TSS per ha. Using the 45° line brings the grower to the production function quadrant. There the grower finds that he needs to irrigate with 1,800 m^3/ha in order to obtain that yield.

DISCUSSION

The tomato processing industry in Israel offers the growers price incentives for quality (in terms of Brix). The existing agrotechnological practices enable the individual grower to control tomato quantity and quality mainly by means of water application.

The form of the biological relationships (production functions) affects the theoretical economic optimum. Therefore, it is important to find out the precise relationship between water and yield and between yield and Brix. The establishment of such relationships is essential to create a better response from the growers to changes in input-output prices, and to a more efficient resource allocation.

The Israeli growers and the industry have adopted a quality incentive price system. This system, if properly constructed, can be used by the industry to achieve a desirable quality and to control produced quantity. These can be obtained such that both manufacturers' and growers' interests are satisfied and the welfare is improved. Moreover, an appropriate quality incentive price system can be an efficient tool for water saving, and therefore provides better opportunities for the individual farmer and for the national economy as well.

REFERENCES

Agami, M. and Eshet, Y., (1985), *Budgets for Processing Tomatoes*, Ministry of Agriculture, Western Galilee District Extension Office, Unpublished mimeo (in Hebrew).

Ben David, Y., (1985), *The Tomato Industry, The World Market and the Competing Countries. Israel Ministry of Agriculture and the Jewish Agency*, The Planning Authority (in Hebrew).

California Tomato Growers Association, Inc., (1986), *Negotiated Prices*, P.O.Box 7398, Stockton, Ca.

Garoyan, L., (1984), *The Processing Tomato Industry of Israel*, Memo.

Geizenberg, H. and Etiel, E., (1987), *How to Grow Tomatoes for Processing*, The Extension Service, Israel Ministry of Agriculture (in Hebrew).

Golan, A. and Shalit, H., (1985), *Using Wine Quality Differential in Grapes Pricing*, Working Paper No. 8508, The Center for Agricultural Economic Research, The Hebrew University of Jerusalem, Rehovot, P.O.Box 12, Israel.

Gorman, W.M., (1980), 'A possible procedure for analyzing quality differentials in the egg market', *Review of Economic Studies*, 47: 843 - 856.

Griliches, Z., ed., (1971), *Price, Indexes and Quality Change*, Harvard University Press, Cambridge, Massachusetts.

Institute of Farm Income Research (IFIR), (1981), *Profitability of Processing Tomatoes in 1983* (in Hebrew with abstract in English).

Israel Ministry of Agriculture and the Jewish Agency, *Factory Costs for Concentrates Processed Tomatoes*, Mimeo, 1985 (in Hebrew).

Israel Vegetables Production and Marketing Board (IVPMB), (1985), *Annual Report on Vegetables and Processing Tomatoes in 1985* (in Hebrew).

Logan, S.H., (1984), *An Annual Planning Model for Food Processing: An Example of the Tomato Industry*, Giannini Foundation Research Report No. 332, University of California.

Luchinsky, U., (1986), *Processing Tomato Response to Water Tensions in Two Climatic Regions*, M.Sc. Thesis, The Hebrew University of Jerusalem, 180 pgs. (in Hebrew).

Rudich, Y., Mazor, I., Gerah, G. and Geisenberg, H., (1985), *Studies in Processing Tomatoes: Mutual Dependency of Quality and Yield and Methods for Monitoring them*, Research Report for the Years 1980 - 1984, Hebrew University of Jerusalem, Faculty of Agriculture, Department of Field Crops, Rehovot 76100, Israel (in Hebrew).

Index